Construction Sealants and Adhesives

2ND EDITION

Julian R. Panek
Consultant

John Philip Cook
University of Cincinnati

A Wiley-Interscience Publication

John Wiley & Sons
New York • Chichester • Brisbane • Toronto • Singapore

Library of Congress Cataloging in Publication Data:

Panek, Julian R.
  Construction sealants and adhesives.

  (Wiley series of practical construction guides)
  Rev. ed. of: Construction sealants and adhesives /
John Philip Cook. 1970.
  "A Wiley-Interscience publication."
  Bibliography: p.
  Includes index.
  1. Sealing compounds. 2. Adhesives. I. Cook,
John Philip. II. Title. III. Series.

TP988.P38 1984   624.1'899   83-21677
ISBN 0-471-09360-2

Printed in the United States of America

10 9 8 7 6 5 4 3 2 1

Series Preface

The Wiley Series of Practical Construction Guides provides the working constructor with up-to-date information that can help to increase the job profit margin. These guidebooks, which are scaled mainly for practice, but include the necessary theory and design, should aid a construction contractor in approaching work problems with more knowledgeable confidence. The guides should be useful also to engineers, architects, planners, specification writers, project managers, superintendents, materials and equipment manufacturers and, the source of all these callings, instructors and their students.

Construction in the United States alone will reach $250 billion a year in the early 1980s. In all nations, the business of building will continue to grow at a phenomenal rate, because the population proliferation demands new living, working, and recreational facilities. This construction will have to be more substantial, thus demanding a more professional performance from the contractor. Before science and technology had seriously affected the ideas, job plans, financing, and erection of structures, most contractors developed their know-how by field trial-and-error. Wheels, small and large, were constantly being reinvented in all sectors, because there was no interchange of knowledge. The current complexity of construction, even in more rural areas, has revealed a clear need for more proficient, professional methods and tools in both practice and learning.

Because construction is highly competitive, some practical technology is necessarily proprietary. But most practical day-to-day problems are common to the whole construction industry. These are the subjects for the Wiley Practical Construction Guides.

M. D. MORRIS, P.E.

v

Preface

This book represents a complete revision of the first edition written by John Cook. Many of his basic concepts still hold true and will continue irrespective of time, since the parameters that affect joint movement have not changed. The once-dominant polysulfides have been replaced by improved urethane and silicone sealants. Innumerable glazing tapes and gaskets are now available to simplify the installation of glass at lower labor costs. Stopless glazing based on structural silicone sealant is now quite common, but demands extreme control on metal surfaces, both in design and installation. The technology on highway and bridge sealants and gaskets has gone through major changes. Hot-melt PVC–coal tar compounds are outperforming past materials on airport runways, and new compression seal and strip seal designs have met the demand for greater compression set resistance and performance on bridge joints.

Sealants are now characterized by joint movement capability that varies from ±5% or less for the oil-base caulks; to ±12.5% for the solvent-base caulks, latex caulks, and butyls; to ±25% for the polysulfide, urethane, and silicone sealants; and up to the latest proposed class of ±50% for the low-modulus silicone sealants and possibly modified urethanes.

Since the first edition of this book, over 50 standards have been issued on sealants, tapes, gaskets, and membranes by ASTM Committee C-24 on Building Sealants and Seals. These standards cover test methods, specifications, and practices, and much of the pertinent information is reflected in comparisons between old and current specifications. Recommendations have been made for upgrading existing standards as well as proposed performance requirements where standards still do not exist.

Architects, engineers, contractors, and students will find this book extremely useful in covering the latest technology on sealants, tapes, gaskets, and other materials, and a sound reference to all specifications on construction sealants and adhesives.

Julian R. Panek
John P. Cook

Yardley, Pennsylvania
Cincinnati, Ohio
November 1983

Contents

*Construction
Sealants and Adhesives*

1

Introduction
to Sealants

The early 1950s saw the application of the new concept of curtain-wall construction in high-rise structures throughout the world. This concept was based on a skeleton of steel or concrete that was then wrapped in a separate envelope or nonbearing curtain wall. This reduced the use of the low-rise bearing-wall-type construction, and introduced more flexibility in design through the use of precast concrete panels, glass or metal panels, unit masonry, and combinations. These segments could either be attached to separate floors or erected on a grid system attached to the structure. The segments could be completely factory-finished and installed on the job site, or erected partially assembled and the remaining units such as window inserts or panels installed later. These structures are inherently more flexible than bearing-wall construction, and present new problems in weatherproofing the structure.

The introduction of the curtain wall also saw the introduction of elastomeric sealants that would adhere to the various surfaces and take greater movement in expansion joints. The old oil-base caulking compounds no longer qualified, and the new materials became "sealants" rather than "caulks." However, the term "sealant" has been too broadly used, and now includes a wide assortment of weatherproofing joint materials. Sealants or caulking compounds include viscous liquids, mastics, pastes, tapes, and gaskets, and the materials can either cure to a rubber or remain in a mastic stage; gaskets and tapes can be supplied in a permanently soft composition or in a semicured or cured state to include various gaskets now widely used in the glazing industry.

The layman's definition for "sealant" is any material placed in a joint opening, generally for the purpose of weatherproofing the building, so designed to prevent the passage of moisture, air, dust, and heat through all the joints and seams in the structure. The ASTM Committee C-24 definition for "sealant" is "In building construction, a material that has the necessary adhesive and cohesive properties to form a seal."

Sealing materials take on a variety of different forms. They may be very fluid and be pumped from an oil can into very thin cracks in order to penetrate the opening and then eventually cure to form a seal. More viscous sealants may be pourable for use in horizontal joints on a sidewalk, roof deck, or parking garage, and their pourability eliminates the need for tooling while giving better wetting to the substrate. Gun-applied sealants or mastics are made thixotropic so that they can be applied to vertical joints without any sag before cure. Gun-applied sealants are easier to apply in narrow joints up to 1 inch in width although tooling is always recommended to force out air bubbles, give better wetting to the sides, and also to smooth the surface for optimum appearance. Tools may be a spatula, a tongue depressor, or pieces of hard wood or metal shaped to the desired width.

Tapes—which may be uncured, partially cured, or completely cured—are used as bedding compounds, for architectural glazing, and in a multitude of areas where a controlled thickness is desired with some structure to support building elements if necessary. The preformed shapes make it easy to furnish tapes in roll form where preset dimensions are required. The tapes can be spliced on the job, and their use enables glazing of the structure from the inside, which results in a savings of time and labor. In some instances, cap beads of a sealant are used where the openings are large, in order to prevent eventual extrusion of the tape during expansion movement of the glass. Silicone sealants are now being used for the larger openings, but solvent-release acrylic sealants are adequate for the smaller units.

Cured gaskets are supplied either in a solid consistency, or porous in a wide range of densities, and are available in a wide variety of shapes for various glazing and sealing jobs. The ideal situation is one in which the exterior gasket is a preformed closed-cell neoprene sponge that locks into a nub on the exterior window flange, while the interior gasket is made using a dense neoprene or EPDM compound that is wedge-shaped and forced into the space between the interior stop and the glass to exert pressure and make a tight fit (see Figure 16.4). The interior gasket is also shaped to lock into a nub on the interior stop. Various curtain wall companies supply their own designs for the wall and gaskets they use. Combinations of gaskets and sealants may be used where all stops are plain in order to get a watertight seal. Although gaskets are more expensive than tapes or sealants, they permit fast erection and also give an excellent finished appearance of the sealed joint.

1.1. The Market

Total new construction is generally 10% of gross national product (GNP). Of this total construction volume, approximately 30% is devoted to heavy construction (bridges, highways, and dams) and 70% is devoted to building construction. Of the building construction volume, 55 to 60% is residential construction and 40 to 45% is nonresidential construction.

New construction generally grows at a rate of approximately 4% per year, but sales in sealant products have grown at approximately 8% per year and are expected to maintain this trend. In addition to the new construction market, the total construction for maintenance, repair, and remedial work in 1982 amounted to approximately $55 billion. Of this, $27 billion represents nonresidential work and $28 billion represents residential work.

Taken together, new construction and maintenance work in 1982 consumed 5 million gallons of sealants valued at $75 million for the commercial market, and 8 million gallons valued at $120 million for the largely residential noncommercial market. The latter area includes the over-the-counter trade for the do-it-yourself market.

While polysulfide sealant at one time was the dominant building sealant, this position has changed considerably. The percentage volumes of sealants types used in the construction industry in 1982 are as follows:

Sealant Type	% Volume Range
Polysulfide	3 to 5%
Urethane	7 to 9%
Silicone	7 to 9%
Solvent acrylic	7 to 9%
Latex acrylic	16 to 20%
Butyl and polybutene tapes	16 to 20%
Solvent butyl	16 to 20%
Oleoesinous materials	8 to 12%
Others	5 to 7%

At the present time use of silicone sealants is expanding in the construction market at the expense of other sealants. The more recent low-modulus silicone sealant is being used more in metal curtain-wall joints because of its greater movement capability. Silicone is also going into the glazing areas. Stopless glazing, another new architectural innovation, requires high-modulus structural-grade silicone sealant.

Butyl sealant is used both in the over-the-counter trade as well as the commercial market, along with asphalts, neoprenes, and oleoresin caulks.

Latex acrylic sealants are also sold to both areas. Solvent acrylic sealant has greater volume in the commercial area, where it has much better quality, but cheaper grades are made for the over-the-counter market. Generally materials for the do-it-yourself (DIY) market are much poorer in quality, since they are sold on a price basis; since homes do not have too much movement, however, these materials might suffice. A small quantity of high-quality silicone sealant is used by the DIY market for bathtub caulk and around ceramic tile and fixtures, where it is one of the best materials and probably one of the few that will perform.

Polysulfide sealant is especially formulated for the insulating glass market, where it still is the major polymer, but both hot-melt butyl compounds and silicone sealant are making inroads. Silicone is especially desired for maximum performance in a silicone/butyl combination, which is preferable when price is not a major concern and the greatest longevity in performance is a must.

In general, these percentages will not vary except where noted. Urethane sealants have shown better resistance to aging, ozone, and ultraviolet exposure, and with silicone sealants will remain the dominant high-quality-performance building sealants.

1.2. Uses for Sealants

There are so many different applications for sealants in construction (see Figure 1.1) that it would be impossible to list them all. It is possible, however, to list general categories of sealant applications and give a few examples in each group.

In nonresidential construction, sealant and caulk applications can be broken down into three general categories; exterior sight-exposed applications (70%), interior sight-exposed applications (15%), and interior and exterior concealed applications (15%). Each of these categories can be broken down into high-volume and low-volume areas.

Exterior Sight-Exposed High-Volume Applications

1. Horizontal and vertical metal-to-metal and masonry-to-masonry expansion and control joints.
2. Dissimilar material joints, such as metal-to-masonry or concrete-to-wood.
3. Joints between precast concrete facing panels.
4. Perimeter of wall and roof openings, such as windows, doors, louvers, pipes, vents and ducts, and chimneys.
5. Pointing of brick and stone masonry.
6. Horizontal paving or traffic-bearing joints as in sidewalks, patios, roof decks, terraces, highway pavements, and expansion joints in bridges.

TYPICAL USES

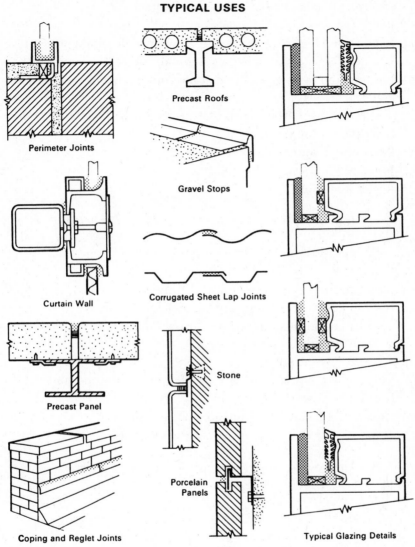

Perimeter Joints

Precast Roofs

Gravel Stops

Curtain Wall

Corrugated Sheet Lap Joints

Precast Panel

Stone

Coping and Reglet Joints

Porcelain Panels

Typical Glazing Details

FIGURE 1.1. Typical sealant applications. (Courtesy of Tremco, Inc.)

7. Channel and face glazing operations.
8. Corrugated metal roofing and walls.
9. Spandrels.

Exterior Sight-Exposed Low-Volume Applications

1. Junction joints, such as the joint where the sidewalk joins the base of the building or where the roof and the side wall of the building meet.

2. Exterior thresholds, flashings, and copings.
3. Pools and reservoirs.
4. Repair caulking of small glazing joints (needle caulking).
5. Between the secondary framing (muntin) and each curtain-wall panel.
6. Between the main vertical metal post (mullion) and each curtain-wall panel.
7. Tappings (waterproofing of bolts, rivets, and other fasteners).
8. Glass panels for skylights.
9. Pipes, ducts, and vent openings.
10. Joints in parapets.
11. Reglet joints.

Interior Sight-Exposed Applications (All Small Volume)

1. Perimeter of doors and fixed window frames.
2. Exposed interior masonry control joints.
3. Joints and recesses between metal frames and interior masonry.
4. Interior construction joints in concrete and plaster.
5. Joints at interior masonry or plaster walls that adjoin columns, pilasters, or exterior walls.
6. Wood-to-masonry joints.
7. Joints where a precast concrete ceiling meets a wood partition.
8. Interior duct work.
9. Plaster and metal trim joints in curtain wall.
10. Expansion or control joints in industrial floors.

Interior Concealed Applications (All Small Volume)

1. Joints in exterior walls.
2. Concealed masonry-to-floor structure joints.
3. Exterior and interior metal thresholds, saddles, and sills.
4. Joints and recesses between access panels, electric panels, piping, and pipe sleeves.
5. Between metal frames and masonry substrates as in show windows.
6. Between overlaps of mating or interlocking metal joints in curtain-wall construction.
7. Between mullions and curtain-wall panels.

In the residential market, builders and painting contractors use sealants for perimeters of wall and roof openings such as doors, windows, chimneys, and vent stacks, and for glazing work. The homeowner uses caulks and sealants for these same applications as well as for filling cracks and sealing tubs and shower stalls. Approximately 80 percent of residential uses are exterior.

1.3. Sealant Suppliers

This industry is characterized by raw-material suppliers that furnish basic ingredients to a large number of formulators. These raw-material suppliers include many of the major chemical companies in the United States. They supply basic polymers or oils to approximately 100 medium to large for-mulators covering a broad geographical range, and to a much more nu-merous group of small companies that operate on a regional basis with fewer products. Several large companies are international in scope and have manufacturing facilities in other parts of the world. Many manufacturers have affiliates in other parts of the world. Table 1.1 gives some information on several of the polymer suppliers.

Although urethane polymer can be bought, many sealant manufacturers make their own polymers from raw materials—which permits them to tailor-make products to suit their own requirements, at lower costs. Among these companies are Tremco, Products Research, H. S. Peterson, and Mameco.

Accessory ingredients include reinforcing fillers, bulk fillers, plasticizers, catalysts, curing agents, activators, retarders, antioxidants, and ultraviolet light absorbers. These are furnished by a wide range of suppliers.

The result of this dispersal is that the quality control of sealant products

TABLE 1.1. Polymer Suppliers

Polymer	Supplier
Polysulfide	Thiokol Chemical (polysulfide polymer is also made in Japan and East Germany for some of the foreign markets and competes with Thiokol there.
Polymercaptan	Products Research makes a special polymer for its own use in an insulating glass sealant. Diamond Shamrock at one time sold both poly-sulfide and polymercaptan polymers but discon-tinued activity.
Silicone	Finished sealants by General Electric, Dow Corn-ing, Stauffer-Wacker, Rhodia, Rhone Poulence.
Silicone	Base polymer now being introduced by Mobay.
Urethane components	Wyandotte, Dupont.
Butyl and polybutene	Polysar, Amoco, Cities Service, Exxon.
Acrylic polymers	Rhom and Haas, Union Carbide.

depends on a multitude of formulators whose expertise would naturally vary quite widely. Another major factor dictating quality depends on the competitive attitude of the formulator and the area where the sealants are sold. Quality-conscious manufacturers may supply sealant to meet various specifications. There is a wide range of prices on the commercial market depending on performance claims, service, and reputation. The over-the-counter trade is highly competitive and controlled by the buying practice of the average home owner. A recent check at several hardware stores revealed that many caulks were selling at a higher price than the price that applicators pay sealant manufacturers for products. However, it must be recognized that the distribution through wholesalers and retailers adds several large profit margins, so that a cartridge selling for $3.00 would probably be sold by the manufacturer for $1.25—which is approximately one half the cost of quality cartridge sealants being sold directly to applicators for commercial projects.

The larger sealant suppliers include Tremco, DAP, Products Research, General Electric, Dow Corning, Protective Treatments, Parr, Pecora, Norton, Mameco, and Woodmont. Larger sealant suppliers probably account for 70% of the sealant market, with the remaining 30% distributed among another 90 or more smaller companies.

1.4. Sealant Specifiers

In the noncommercial or consumer market, the contractor is the primary buying influence on new construction. Approximately 75% of the time contractors specify not only the type of product but also the brand.

In the repair and maintenance market, small painting contractors, siding contractors, and homeowners are the buyers of sealant products.

In heavy construction the engineer specifies what sealants will be used. This type of construction includes highways, bridges, dams, airports, and water supply and sewer construction.

In new nonresidential construction, the sealant and compounds are essentially the choice of the architect. In some cases very reputable architectural firms will write a closed specification on the specific sealant and manufacturer they desire. In the majority of cases, the architect will insist that the sealants meet a performance specification, and will then list two or three manufacturers and their products as well as including a statement "or equivalent." This clause does open the door for applicator to submit either a better or a cheaper product. The applicator's proposals must be approved by the architect, who may not be qualified to judge the quality of the proposed sealant. On most city, state, and federal projects only government specifications or approved equivalents can be used, and applicators are at the mercy of lower-price sealants. Nonresidential construction includes educational, industrial, commercial, religious, institutional, social, recreational, and manufacturing buildings. In addition, the architect is the primary influence in major nonresidential remodeling, and in custom residential homes, which represent about 15% of housing starts.

Manufacturers of certain prefabricated units such as windows and panels use a sizable segment of sealant materials, but these companies exercise their own quality control and normally furnish a warranty on the finished work. Warranties vary considerably, depending on whether the units are to be used for condominiums or for office buildings.

1.5. Specification Methods

Architects or specification writers differ in the pattern they use in specifying sealants. The approach varies according to the size, type of project, owner, and specification policy of the firm. Listed below are some of the approaches used to specify a sealant material:

1. Give a reference standard such as a federal specification, ASTM standard, or ANSI standard for most city, state, or federally funded projects.
2. Specify generic type sealants such as urethane, polysulfide, silicone, or solvent acrylic, and name two or three manufacturers.
3. Cite a reference standard along with one or more approved sealant manufacturers and their specific sealants.
4. Specify a reference standard along with an approved generic class of sealant.
5. Indicate the performance and physical and chemical properties required of the sealant, and name two or three sealants that meet these characteristics.
6. Specify the products of one company only.
7. Name the specific trademark products of several sealant manufacturers.
8. Indicate only the generic classes of sealant such as urethane, polysulfide, silicone, or solvent acrylic.
9. Use the AIA master specification along with its list of sealant manufacturers.

All of the above approaches are used, and the choice is evidently based on company policy. The safest specifications are those in which only one product is specified. In all other cases, the listing of several sealant manufacturers does not always result in the better sealant being used. Many applicators may have their own preferences based on experience. The decision may be to select one of the three or more sealants approved by the architect, or even to submit their own choice for approval by the architect on an "or equivalent" basis. The final choice often depends on who takes responsibility.

In nonbuilding construction—especially bridges and highways—much of the work is funded by the individual authority, which may be city, state, or federal. Most states have complete construction standards and specifications based on federal or ASTM specifications and test methods. In public bidding construction, open bidding somewhat limits architects in their specification writing. In this type of work, the architect generally must include an "or equivalent" clause in the specification, which means that the architect may not limit acceptance to certain proprietary materials. Although the bidder may substitute any materials that meet the performance requirements of the specification, the bidder should obtain the architect's approval of an "or equivalent" material. This generally does not occur—which is evident by the quality of workmanship on much public construction. In private building construction, the contractor can be overruled by the architect.

The architect is therefore the person held responsible for the sealants and caulks he or she specifies. Because of this liability, architects naturally tend to be conservative in their choice and use those materials with which they have had a successful record of performance. While architectural drawings may reflect proper practice, job-site construction may cause joint sizes to vary considerably. If joints occur that are too small for the movement capability of the sealant, the sealant will fail—thus creating problems that require considerable remedial work. Placing the blame is another complex problem. Many architectural firms now have resident architects to oversee all facets of construction in order to minimize these types of errors. Also, some sealant manufacturers work very closely with applicators, particularly before work begins, in order to review job problems. The astute applicator is one who cases the job before bidding. Unfortunately, not all architects can design joints properly even though the state of the art has reached a high level of performance. Many failures that have occured could have been prevented with proper design. On the other hand, for nonspecified private remedial work, the caulking contractors are generally concerned with nonworking joints and are partly justified in using the lowest-priced materials available. However, there is a point of no return—where these sealants shrink, become hard with time, and fall out. It would be a positive development if home builders did give a little more regard to sealant quality in this day of high housing costs.

The architect will frequently hold the sealant manufacturer responsible for the performance of the material, and will require that a manufacturer's representative be present at the job site before and during installation. Moreover, many architectural firms require that the manufacturer offer a performance guarantee for a specified time period and even insist that the manufacturer pay for a complete recaulking job if the building leaks. However, from a sealant manufacturer's standpoint, a good sealant can be misapplied; and the applicator is therefore responsible for the job. Sealant manufacturers have no control over the misuse of their sealants, and if they are consulted on any job will usually warn architects or applicators in writing of the improper conditions at the job site. The ideal situation is for architects to review their drawings with a technical sealant representative. Any design problems can be easily corrected on paper.

In order to consider a sealant for use on a building, architects often insist on complete technical data sheets from the manufacturer. These "tech data" sheets contain information about the following properties of a sealant:

1. Pot life.
2. Cure time.
3. Mixing instructions for multipart sealants.
4. Modulus of elasticity.
5. Ultimate elongation.
6. Hardness before and after heat exposure.
7. Application temperature limitations.
8. Performance temperature limitations.
9. Color and color retention after ultraviolet exposure.
10. Ultraviolet resistance and retention of properties after accelerated exposure.
11. Weight loss after heat aging or long-time aging.
12. Bond durability.
13. Gunability.

The data sheets should also indicate which federal or ASTM specification the material meets. The data sheets may give design criteria and limitations of the sealant's performance, and may contain application instructions and other helpful points that the architect should review. Naturally, these specifications are considered minimum by the industry; hence the data sheets usually indicate to what extent the materials exceed the specifications. However, data sheets are not always accurate, and it would behoove an architect to require certification from an approved testing laboratory that a given sealant meets the major specifications. Although architects have on occasion included extra performance requirements, they are not sufficiently familiar with sealant performance to know which properties or requirements to upgrade. Most architects insist that the supplier furnish the results of roughly three years of field experience before specifying a new material for a major project. The best criterion is past experience with specific sealants, which becomes the basis for many architects.

Representative costs and typical applications for specification-type sealants are given in Table 1.2.

TABLE 1.2. Typical Sealant Applications and Representative Cost

Sealant Base	Cost Range (dollars per gallon)	Uses
NONHARDENING SEALANTS		
Oleo-resin	6–15	Sealing of concrete joints, masonry copings, glazing and sealing of cable splices, and glass-to-metal meter cases.
Asphalt and bituminous	3–8	Sealing faying surface metal joints, silos, and air conditioners. Caulking for expansion and contraction joints.
Butyl	6–15	Caulking expansion and contraction joints, Metal-to-glass seals, and metal-to-metal sealing to separate dissimilar materials. Sealing electrical conduit.
Acrylic	6–15	Pipe joints, glazing, masonry, and metal caulking. Special compounds used as liquid gaskets and pipe dope.
Polybutene	6–15	General construction-type caulking and glazing. Seal between dissimilar metals. Metal curtain wall.
HARDENING SEALANTS—RIGID TYPES		
Epoxy	8–20	Potting electrical connectors, encapsulating miniature components, coating circuit boards, and cable splicing. Caulking and pipe sealing. Used as a sealer and abrasion-resistant coating for concrete. Concrete repair compound.
Polyester	8–20	Potting, molding, and encapsulating. Gasket and pipe thread sealants.

Oleo-resin asphalt and bituminous	3–8	Same general uses as nonhardening formulations. Hardening type materials should be used where pressure limits exceed the limitations of the nonhardening formulations.

HARDENING SEALANTS—NONRIGID TYPES

Chemical-Reaction Systems

Polysulfide—two part	15–25	Sealing integral fuel tanks. Sealing faying surfaces and channels. Potting, molding, and sealing of plexiglass. General construction sealing and caulking of metal, wood, and masonry joints.
Polysulfide—one part	15–25	General construction-type sealing, caulking of metal, wood, and masonry joints.
Urethane—one part	15–25	General construction-type sealing, caulking, and some glazing of metal and wood joints.
Urethane—two part	15–25	Potting and molding of electrical connectors, encapsulation of hydrophones, transducers, and circuit boards. Caulking where compatibility with liquid oxygen is required.
Silicone—one part	15–25	General construction-type sealing, caulking, and glazing. Potting and molding.
Silicone—two part	15–25	Potting and molding of electrical connectors. Potting firewall connectors, and coating umbilical cables. Sealing of heat shields. General construction-type sealing, caulking, and glazing of metal, wood, masonry, and glass-metal joints. Structural glazing.
Polysulfide—modified Bitumens—two part	10–15	Sealing of expansion and contraction joints in runways and taxi strips where flame and fuel resistance is required.

(continued)

Table 1.2 (*Continued*)

Sealant Base	Cost Range (dollars per gallon)	Uses
Modified epoxy—two part	10–20	Potting, molding, encapsulating. Sealing transformers. High-voltage splicing, capacitor sealing. General construction caulking.
Acrylic latex—one part	10–15	Interior joints on wood, masonry, and plaster.
Viton—two part	80–115	For high-temperature service where fuel and oil resistance is required. Sealing fuel tanks and faying surfaces.
Solvent-Release Systems		
Neoprene	10–15	Sealing between dissimilar metals. Caulking and general sealing. Low-movement joints.
Hypalon	10–15	Similar in use to neoprene.
Butyls	10–15	Glazing and caulking of metal, glass and masonry-type joints.
Acrylic	10–15	General construction-type sealing, caulking, and glazing. Low-movement and medium-movement joints.

1.6.   Specifications

In putting together the contract drawings for a structure, the architect writes a set of specifications covering materials and their installation. It is common practice for architects to refer to other reference standards and to include them as a part of the document. For example, the architect's specification might state, "The sealing compound for the joints in the precast wall panels shall be a one-part urethane sealant conforming to the requirements of ASTM Specification C-920." In essence the architect uses certain standard references as guides and may add other requirements to modify the standard specifications to suit the performance requirements of a particular structure.

Unfortunately, when sealants were first introduced architects were not too familiar with sealants, and sealant manufacturers were not familiar with joint design or movements. Since there were no specifications, several sealant manufacturers were convened by the Thiokol Corporation in 1956, and a first specification was developed that eventually became ASA standard A116.1 1960. A more detailed treatment of the development of the early specifications is covered by Panek (1). The ASA standard then served as a model, and the National Bureau of Standards developed several specifications for one-part and multipart sealants that rapidly went through a number of revisions, resulting in several final standards. The standards are as follows:

1. Interim Federal Specification TT-S-00227E (227 E) Sealing Compound; Elastomeric Type, Multi-component (for caulking, sealing, and glazing in buildings and other structures).
2. Interim Federal Specification TT-S-00230C (230 C) Sealing Compound; Elastomeric Type, Single Component (for caulking, sealing, and glazing in buildings and other structures).
3. Interim Federal Specification TT-S-001543 (1543) Sealing Compound; Silicone Rubber Base (for caulking, sealing, and glazing of buildings and structures).
4. Interim Federal Specification TT-S-001657 (1657) Sealing Compound, Single Component, Butyl Based, Solvent Release Type (for buildings and other types of construction).

The silicone standard was probably designed because silicone sealants could not meet the broader requirements of 227E. A close examination reveals that standard 1543 is almost identical to 227E, but made using a slightly modified test assembly in the cycle test. The butyl standard is for relatively nonmoving joints.

Although the standards developed by groups of sealant manufacturers and ASTM members become reference standards by concensus, the standards developed by the NBS did not, and were designed to meet more rigorous demands on the building under more extreme climactic conditions. The federal specifications for dynamic sealants all include the Hockman bond durability test cycle—which is the one test that is highly controversial from a practical point of view, yet is the heart and soul of these specifications. Without this test, the specifications could easily be met by most materials on the market.

The NBS has withdrawn from any further involvement with sealant specifications and has recommended that the ASTM C-920 specification be used in place of 230C, 227E, and 1543. ASTM specification C-920 was derived by ASTM C-24 as a replacement for the upgraded ANSI A116.1 1967. The A116.1 1960 specification had been assigned to ASTM and then to the newly formed ASTM C-24, which was responsible for all future revisions. ASTM C-24 upgraded the specification in 1967, and then eventually derived the C-920 specification, at which time ANSI A116.1 1967 was withdrawn. ASTM Committee C-24 is very active and presently at work on a number of test methods, standard practices, and specifications for all types of sealants, caulks, back-ups, tapes, gaskets, and other products and as of 1982 had approved over 75 standards.

Although there are a number of groups working on sealant and gasket specifications, very few specifications have received enough acceptance to be considered as standards by the sealant industry. In 1968 the National Association of Architectural Metal Manufacturers published several specifications for noncuring compounds such as butyls together with specifications for sealing tapes. These specifications have received some acceptance, and will probably serve as reference material by the Tape Subcommittee in ASTM C-24.

The Canadian government has issued a number of sealant specifications in the last 20 years, many of which reflect the basic tests used in the NBS standards. They will be included in the comparison of specifications.

Interestingly enough, for private construction work there is actually no such thing as "obtaining approval" for a sealant under a federal specification. The federal standard contains certain requirements that the sealant must meet. Hence the manufacturer must obtain copies of the federal standard and either conduct the tests in the manufacturer's own laboratory or have them performed by an independent laboratory. As soon as the manufacturer is satisfied that a material meets the requirements of the federal standard, the manufacturer may so claim in promotional literature.

Most architects are aware of the limitations of the standard specifications, and consequently tend to accept the manufacturers' technical data sheets with a certain degree of caution. The majority of architects will insist on a proven record of field performance before accepting a material for use on a given project. However, a sizable minority of architects feel that accelerated laboratory testing by a reputable, independent laboratory will suffice. Some well-known and established laboratories with a capability in this area are the following:

Vjorksten Research Laboratories, Inc.
Madison, Wisconsin.

Battelle Memorial Institute
Columbus, Ohio.

Pitt Testing Laboratories
Pittsburgh, Pennsylvania.

Foster D. Snell, Inc.
Cambridge, Massachusetts.

United States Testing Co.
Hoboken, New Jersey.

Southwestern Laboratories
Fort Worth, Texas.

D/L Laboratories
New York, New York.

Associated Laboratories, Inc.
Dallas, Texas.

ETL Testing Laboratories, Inc.
Cortland, New York.

D/L Laboratories have been doing considerable testing of sealants against both federal specifications and the ASTM C-920 specification. They have been recently approved by the Canadian Government Specification Board to test sealants against Canadian specifications.

If a manufacturer wishes to sell products directly to the federal government, the entire specification picture changes. The manufacturer must complete a bidder's mailing list application and submit a bid on the particular job. The purchase by the government is generally made from the low bidder. Before the actual award is made, a government inspector may be sent to the plant to insure that the manufacturer has the necessary production facilities. Moreover, for certain products there is a "quality producers list" (QPL), whereby only the bids of certain approved manufacturers will be accepted upon submission. In order to be placed on this list, a company must submit samples of its products for analysis and approval by the federal government. The picture changed considerably in 1982 with the curtailment of various government testing agencies, and the exact status and the future of these testing facilities is not known. Building sealants were evaluated at the National Bureau of Standards Laboratory in Washington, D.C. Highway and airport paving sealants were tested at the Ohio River Division Laboratory of the Corps of Engineers in Cincinnati. Asphaltic materials were evaluated at the Waterways Experimental Station of the Corps of Engineers in Vicksburg, Mississippi. Sealants for canals, dams, and other reclamation projects are handled by the Bureau of Reclamation in Denver, Colorado.

1.7.   Mock-Up Testing and Dynamic Testing

The multitude of surfaces today to which sealants must adhere—such as anodized aluminum finishes, fluorocarbon-treated surfaces, various aluminum alloys, stainless steel alloys, glass finishes, and precast panels contaminated with form oils—have imposed a tremendous problem on the expected adhesion of all sealants. The architect is advised to make samples of these surfaces available for adhesion testing by reliable sealant manufacturers, and in the end to insist on a mock-up trial test of sealant materials on the job site to duplicate all the conditions. Such steps, when conducted by the applicator, a representative from the sealant manufacturer, and the architect, have in many cases resolved problems before they reached a major magnitude. Often the use of a special primer or masonry conditioner has resulted in satisfactory adhesion that would not have been detected in the laboratory due to lot variation in building materials or special conditions on the job site.

Dynamic testing is resorted to more and more by large architectural firms, which send sections of walls to one of several testing centers, such as Construction Research Center in Miami, Antoine and Associates in St. Louis, or the Commonwealth Research Station in Sydney, Australia. At test centers in Tokyo earthquake movements have been simulated. Such testing exposes the entire wall test section to high winds of 135 mph, and on special jobs to 165 mph. A common testing procedure will call for buffeting of the test wall with the wind generator alternating between 100- and 135-mph wind gusts, along with water spray to test wall components and sealant for leaks and movement resistance.

Such testing has resulted in modifying building components, gaskets, seals, and glass sections to better resist the forces of nature. Static testing is often relied on using both pressure and vacuum with simulated rain to determine the soundness and watertightness of glazed and sealed window sections. We have learned that on high buildings, not only does positive wind pressure have an effect on glazing joints, but in addition negative pressures are created at corners or close to roof lines that can suck glass out of building openings. This has brought about the need for advanced technology in developing continuous pre-shimmed tapes for glazing in order to give continuous support to glass around its entire perimeter.

2

Sealant Classification

2.1. Classification Parameters

Sealants can be classified in a number of ways, and in many cases several classification parameters are needed to understand the essential features of each sealant. Several obvious parameters by which sealants can be classified for specification purposes are the following:

1. Generic type.
2. Movement capability.
3. Cure mechanism.
4. Recovery properties.
5. Performance standards.

2.2. Generic Types of Sealants

One reason for the large number of generic types of sealants is cost. It is apparent that a good silicone or urethane sealant could perform satisfactorily in most sealant application areas, but the high performance characteristics would not necessarily be needed. The sealant cost would be prohibitive in a competitive environment. Specifically, an oil-and-resin-based sealant does a satisfactory job on small metal sash, at a considerably lower cost. Some solvent-release sealants give good adhesion to most surfaces and perform satisfactorily in joints with smaller movement. The solvent-release sealants vary considerably in cost, depending on the areas of application and the solvent content. Many sealants are designed for a one-use appli-

TABLE 2.1. Sealant Classification

Generic Type	Movement (±%)	Recovery	Cure Type	Standards Federal	ASTM	Canadian
Oil-base	2	poor	oxidation	TT-G-410E	none	19GP-1G, 19GP-6a
Oil-and-resin-base	5	poor/fair	oxidation	TT-C-598	C-570	19GP-2b
Butyl-mastic	5	poor	solvent release	none	none	none
Butyl-curable	7.5	poor/fair	solvent release	TT-S-001657	none	19GP-14M
Butyl-polyisobutylene	10	poor	solvent release	none	none	none
Polyisobutylene	10	poor	thermoplastic	none	none	none
Solvent acrylic	12.5	fair	solvent release	TT-S-00230a	none	19GP-5m
Emulsion acrylic	10	poor/fair	water release	none	C-837	19GP-17
Chlorosulfonated-polyethylene	12.5	fair	solvent release	none	none	19GP-26
One-part polysulfide	12.5	fair	moisture catalyzed	TT-S-00230C	C-920	19GP-25
Two-part polysulfide	25	fair/good	catalyst cure	TT-S-00227E	C-920	19GP-3M
One-part urethane	25	good	moisture catalyzed	TT-S-00230C	C-920	19GP-16a
Multipart urethane	25	excellent	catalyst cure	TT-S-00227E	C-920	19GP-15M
Silicone-structural	25	excellent	moisture catalyzed	TT-S-001543A	C-920	19GP-9M
Silicone-low modulus	40	good	moisture catalyzed	TT-S-001543A	C-920	19GP-9Ma
Nongeneric	25	good	catalyst cure	TT-S-00227E	C-920	19GP-24

cation. Specific areas of application are discussed in the various chapters on generic types of sealants.

The various generic types are shown in Table 2.1 along with other pertinent classification data.

2.3. Movement Capability

One of the most comprehensive classification methods is by movement capability. Movement capability is expressed as a ± % value, with the plus value the amount of movement that the sealant can take in extension in a typical joint, and the minus value the amount of movement that the sealant can take in compression in the same joint. In each case the comparison is to the original dimension of the joint at the time the sealant is installed. Thus, a sealant with a ±25% joint movement capability can take 25% in movement in either extension or compression during the lifetime of the sealant based on its original joint dimension at the time of installation.

Most sealant standards today include a qualifying statement that specifies the joint movement capability of sealants covered by the standard. Some standards may include several classes, such as Class 12.5 and Class 25 in ASTM C-920 for ±12.5% and ±25% joint movement.

It is readily apparent that sealants that set to a tough, inflexible compound, such as oil-based sealants, have low movement capability. The various solvent-release sealants have movement capability up to ±12.5% since they exhibit thermoplastic properties. Sealants that cure to a rubber-like elastic compound are the sealants with movement capability of ±25% or more.

From an architectural standpoint, the movement capability is the best parameter for qualifying sealants for various areas of application.

There are new low-modulus silicone sealants on the market today with claims of ±50% movement capability, which gives greater latitude for joint design. ASTM standards do not exist for this class, but ASTM Committee C-24 is considering this class along with appropriate test methods and requirements.

2.4. Cure Mechanism

Solvent-release sealants are essentially thermoplastic, since the sealant is made by incorporating a good quantity of solvent into a gummy synthetic rubber. For all practical purposes, very little cross-linking takes place, and the sealant in essence converts to the original rubber upon evaporation of the solvent. The addition of various fillers gives the sealant more structure.

The oil-based sealants are cured by air oxidation, and the straight linseed-oil-base sealants convert to a tough, relatively inelastic compound. Oil-and-resin-based sealants for use on metal sash are modified with various non-reactive oils so that the final sealant retains some elastic properties while slowly curing from the outside to the inside.

It is necessary for the base polymers to react with all the terminal groups and exhibit some cross-linking in order to obtain more elastic properties. The curing agents may be incorporated in a separate component, or may be included in the base sealant for one-component sealants. These one-component sealants cure by the catalytic action of moisture from the air in a short time. Silicone sealants can be completely cured in 24 hours. Other sealants may require several days for the mechanism to give a complete cure. These more complicated polymeric systems are capable of greater movement, since the cured polymer structure is three-dimensional.

2.5. Recovery Properties

The solvent-release sealants all exhibit poor recovery. For this reason the movement capability is limited to $\pm 12.5\%$. Greater movement would result in a permanent distortion of the sealant in the joint, which would eventually result in failure. The catalyzed sealants all exhibit better recovery, and consequently can be used in joints having greater movement. The urethane sealants exhibit better recovery than the polysulfide sealants, and the best recovery is shown by the structural silicone sealants. These sealants can be used in "stopless glazing" where the silicone sealant is the only means of holding the glass in place. The sealant then becomes a structural sealant, and hence the application is called "stopless glazing." The recovery of this silicone sealant is phenomenal, and sealants will exhibit over 95% recovery after being compressed for one year at elevated temperatures. No other generic class of sealant can exhibit this excellent recovery.

2.6. Performance Standards

Architects refer to sealant standards in their architectural specifications to describe their choice of sealants for various application areas. The most widely used federal standards are TT-S-00230C, TT-S-00227E, and TT-S-001543A. These are being replaced by ASTM C-920, which is nongeneric and covers both one-component and two-component sealants. The ASTM C-920 standard is given in Appendix 1. This standard has shown in recent years that some of the requirements need upgrading. Recommendations are given in Appendix 2 for these upgraded revisions, and architects would benefit by the use of these recommendations.

Other standards are listed in Table 2.1 for the various generic types of sealants. The Canadian standards listed are useful and in many cases also reference ASTM test methods.

Federal, state, and city projects require that sealants can only be specified by standards—specific company products cannot be listed. Yet in most cases the best method of specifying a sealant is by a closed specification. This reflects personal choice and past experience. Some architects have a policy of listing several products as well as the expression "or equivalent," which essentially opens the door to the least expensive product. A closed specification is the best protection for the architect, and discriminating architectural firms also supervise much of the sealant installation.

3

Weatherproofing the Building

3.1. The Need for Sealant Materials

The exterior of a building must be weatherproofed in order to prevent drafts and wind-driven rain from entering the building. With today's emphasis on year-round air conditioning, it is of paramount importance that the building be properly sealed in order to maintain this level of conditioned comfort. Water leakage into a structure may cause structural deterioration and usually presents unsightly damage and costly repairs. Far less dramatic, but just as costly in the long run, are the increased costs of heating and air conditioning that can result from a poorly sealed structure.

Most buildings are well-designed and properly sealed when they are new. However, movement of the building or of its component parts often causes sealant failure and resulting leaks. Movement of the building might be caused by expansion and contraction of its components, wind load on the structure, or settlement. Buildings of heavy bearing-wall construction tend to remain fixed and absorb any movement over a large number of joints. In the heavy brick wall, for instance, building movement may show up as fine hairline cracks in the mortar joints, which are not, in general, too susceptible to rain damage. In time, however, with the intrusion of moisture, alternate freezing and thawing can open up these joints and cause spalling. In many cases the mortar is porous and the surface wall may require treatment of a water-repellant material that still permits breathing. On the other hand, the large, flexible structure with relatively large panels—characteristic of today's curtain-wall construction—may move a great deal. The joints between large precast panels may move ¼-inch or more due to expansion and contraction. During a high, gusty wind, the top of a large, flexible curtain-wall building may sway several feet out of plumb. Naturally, such severe movements place a great deal of strain on any joint-sealing material.

3.2. The Mechanism of Rain Penetration

In order for rain to penetrate the wall of a structure, there must be water impinging on the wall of the building, an opening in the wall through which the water can travel, and a force to drive the water through the opening.

During a driving rainstorm the water forms a film on the exterior surface of the building. This film of water flows down and over the joints in the building surface. If there are any openings in the joints, the wind pressure acting on the surface forces the water into or through the building wall. Another factor is that the building may have a lower-than-normal atmospheric pressure inside either due to ventilation equipment within the building or wind movement around the building—either of which can suck water into the joint. Estimates vary as to how much water flows down a vertical wall. One research institute uses a figure of 40 gallons per 100 square feet per hour. Against an even wall surface the film of water maintains a rather constant thickness. However, projecting mullions tends to split the air flow and cause concentrations of water flow at the panel-to-mullion joint. Other projections on the building, such as cornices, also tend to cause turbulent air flow and water concentrations. One has only to walk around the streets of a city to see dirt stains distributed on the surface by water flow. The absence of dirt indicates a high velocity of flow.

Other forces that can cause leakage are gravity, capillary action, surface tension, and kinetic energy of wind or wind-driven rain. Leakage due to gravity usually occurs when some part of the roof deteriorates to form an opening. Leakage may also occur along ledges and balconies where sealant may have hardened and shrunk. Under the influence of wind, raindrops approach the wall with considerable velocity, and their momentum alone can carry them through openings of sufficient size. Cover battens and internal baffles can be used to prevent leakage by this method. Surface tension can cause water to flow along soffit edges. This can easily be prevented by the use of a drip at the outer edge of an overhang. Capillary action can occur whenever two wettable surfaces are close together. This can be solved by introducing an air gap or cavity. The conventional approach to solving leakage is to eliminate all openings by the use of sealant. The more effective and more reliable approach is to eliminate the pressure differential across the opening by equalizing the pressure on opposite sides of the opening. This approach is known as the "rain-screen principle."

3.3.   The Design of Exterior Wall Joints

There are two basic methods of sealing the joints in the exterior wall of a building:

1. One-stage weatherproofing, which is the sealing of joints at the exterior face of the wall section (the wetted plane). Here the sealant material acts both as seal and air seal.
2. Two-stage weatherproofing, in which the air seal and rain seal are treated as separate functions. A deflector or rain shield is used to keep water out of the joint, and the air seal is placed in an interior, protected location.

Although each system of sealing has its advantages and disadvantages, any through-the-wall joint requires some form of seal. Since the functions of the seals are different in the two systems, the performance requirements of the seal also vary.

3.3.1.   One-Stage Sealing

Until very recently most American building practice has been based on the principle of one-stage sealing. Most repair and resealing work is also necessarily one-stage sealing. In addition, many sealant applications such as the perimeter of wall and roof openings for louvers, vents and stacks, junction joints at the base of building, lap joints in sheet materials, and horizontal traffic joints, provide no other solution.

One-stage sealing is simply the elimination of all openings that would admit water past the exterior wall line of the building. In this method of sealing or waterproofing, the caulking contractor installs the sealant in the joint, using standard caulking methods. Thus, familiarity favors the one-stage seal. Figure 3.1 shows a typical one-stage seal and its installation.

The economics of sealing supports the one-stage seal simply because it is a one-step process. Another economic factor to be considered is the design of the edges of exterior building panels. The edges of panels that form the interface of the joint can be of simple straight-line design. (See Figure 4.1).

One-stage sealing, however, has several disadvantages that must also be considered. The sealant material is exposed to the rays of the sun, which may cause damage to the sealant by ultraviolet radiation; the sealant and its substrate are subject to alternate wetting and drying; extension of the exposed sealant occurs when the building contracts in cold weather, and most sealants are less extensible when cold; also, the exposed sealant is subject to picking and gouging by vandals. However, movement is not a problem if the joint has been designed around the movement capability of the sealant selected for the job. There are sealants that will and have with-

FIGURE 3.1.   One-stage sealing. (a) Placing the back-up material. (b) Placing the sealant. (Courtesy of Tremco, Inc.)

stood the effects of the weather, but proper specification, selection, and application are needed to insure good performance.

3.3.2.   Two-Stage Sealing

Two-stage sealing has gained widespread acceptance in Europe and Canada, and has been used minimally by architects in the United States. The foreign growth was partly due to merit and partly to economics. The higher-priced sealants that have been used widely in this country are not as readily available in many countries. Consequently, the Europeans and Canadians have developed joint designs that permit the use of lower performance (and price) materials. However, the use of the curtain wall in modern rain-screen design has also introduced greater movement, so that high-quality sealants are imperative from a movement capability standpoint.

Two-stage sealing treats the rain seal and the air seal as separate functions. The rain seal in horizontal joints can be achieved by a deflector or an offset in the edge of the building panel (Figure 3.2). In vertical joints, the rain shield can be a simple strip or tube of rubber placed in a recess in the edge of the panel (Figure 3.3). The possibilities for forming this rain seal are endless once the designer recognizes that the seal should not be airtight. Several ways in which the rain seal can be achieved in horizontal joints are the following:

1.   The seal can be made using a sealant that is vented on 5-foot centers.
2.   The seal can be designed as a ship lap joint.
3.   The seal can be fashioned as a gasket that fits into reglets in the sides of both panels.

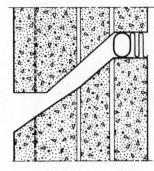

FIGURE 3.2.   Horizontal joint—offset panel edge with rain seal. (Joint design by Trygve Isaksen, Norwegian Building Research Institute.)

Two-stage sealing depends for its effectiveness not on completely preventing water penetration, but rather on controlling the forces that act to drive the water inward. Its function is to prevent water from reaching the interior seal and thus from reaching the inside of the building. The important feature of two-stage sealing is the presence of a pressure-equalizing chamber behind the rain seal (see Figure 3.4). This pressure-equalizing chamber must be vented to the outside so that the air pressure in this chamber is the same as the external air pressure. Thus any wind-driven rain that does pass the rain baffle is trapped in this enlarged area and is drained harmlessly back to the exterior. The two-stage, or rain-screen, principle of weatherproofing has been around for centuries in the form of the traditional shingled wall (Figure 3.5). The rain-screen shingle dispels the force of wind-driven rain, deflecting it away from wall sheathing. Overlapping shingles prevent water entry by gravity, and the spaces between shingles are too large to draw in moisture by capillary action. American home builders have also been using two-stage sealing in their design. The general rule for insulating a residence is to insulate the inhabited areas and ventilate the rest of the structure. Thus the roof serves as the rain shield, attic areas are ventilated, and the air seal (insulation) is placed in a protected region between the ceiling joists.

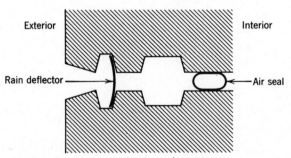

FIGURE 3.3.   Vertical joint with rain seal and air seal. (Joint design by Trygve Isaksen, Norwegian Building Institute.)

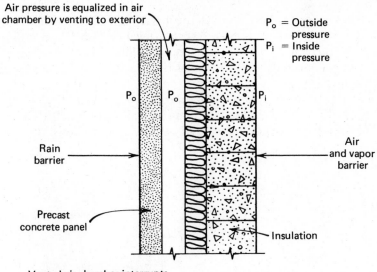

Air pressure is equalized in air
chamber by venting to exterior

P_o = Outside
      pressure
P_i = Inside
      pressure

P_o P_o P_i

Rain
barrier

Air
and vapor
barrier

Precast
concrete panel

Insulation

Vented air chamber interrupts
water passage through outer
cladding and keeps inner wall dry

FIGURE 3.4. Rain-screen principle applied to wall with precast concrete facing. Air chamber between rain-barrier panel and interior air seal is vented to outside to equalize pressure and facilitate drainage.

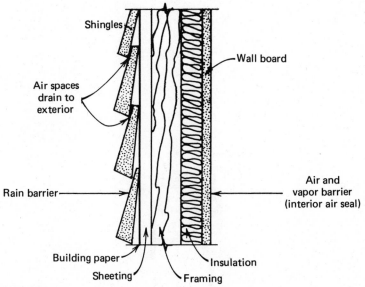

Shingles

Wall board

Air spaces
drain to
exterior

Rain barrier

Air and
vapor barrier
(interior air seal)

Building paper

Insulation

Sheeting

Framing

FIGURE 3.5. Rain-screen principle applied to shingle wall. Shingles deflect rain away from wall sheathing and prevent water entry by gravity. Interior wall finish acts as interior air seal.

29

Two-stage sealing therefore has the advantage of having the sealant material (the air seal) placed in a protected location. Thus the sealant is not exposed to ultraviolet radiation, extreme cold, or a wet-dry cycle. Under these circumstances the service life of the sealant may be substantially extended.

There has been some differentiation between types of two-stage joints. One is classified as the elementary two-stage joints in which all interior and exterior joints are sealed and vented at joint intersections. The modified

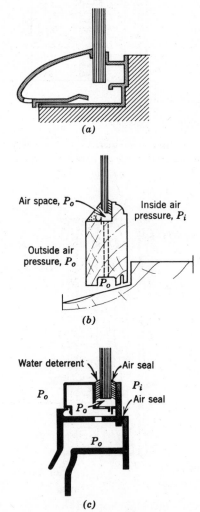

FIGURE 3.6.   (a) "Store front" glazing detail showing openings for air-pressure equalization and drainage. (b) Single glass in wooden sash with openings for pressure equalization and drainage at sill. (c) Metal sash unit designed for pressure equalization. (Design by G. K. Garden, Canadian National Research Council.)

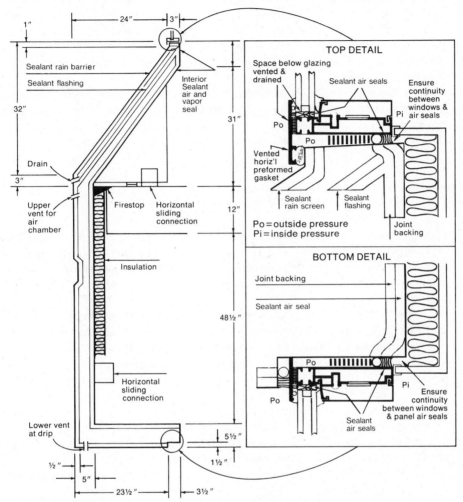

FIGURE 3.7.   A rain screen window detail. (Courtesy of Tremco, Inc.)

two-stage joint provides better weather protection by introducing a sizable air chamber between both walls and interrupting the air chamber by flashing or sealant usually at the top of a panel, to prevent chimney action in the cavity, and modifying the panel edge configuration to design an offset joint to provide a rain barrier for horizontal joints.

Two-stage sealing has been applied to glazing units. Store-front glazing details allow for openings for pressure equalization and drainage. Many buildings now use a dry/dry method for glazing either single or insulating glass panels in which compression gaskets are used both on the inside and outside surface of the glass. The air space beneath the glass units is vented to the outside. Other configurations for glazing are classified as wet/wet and wet/dry, in which case "wet" signifies the application of a curable sealant (see Figures 3.6 and 3.7). A silicone sealant may be applied over a glazing gasket which by curing adheres to both glass and metal and insures that the glazing gasket will not work out with expansion and contraction of the glass with time. The sash design dictates the selection of glazing components that may be used.

The two-stage seals also have some disadvantages. The sealant material is generally more difficult to apply. Resealing in the event of failure may be extremely difficult. The designer must also consider the cost of forming the necessary recesses of offsets in the panel edges to accommodate the rain baffle and the pressure chamber.

3.4. Types of Joints

From a functional viewpoint, there are only two types of joints in building and construction: (1) working joints, which change size or shape with the relative movement of the adjoining parts; and (2) nonworking joints, in which relative movement between adjacent parts is minimized or eliminated by the details of construction.

In low-rise buildings or buildings that cover large areas, the major working joints are isolation joints and control joints. The isolation joint, which is approximately 1 inch wide, divides the building into entirely separate sections. It is a through-the-building joint and is spaced at intervals of approximately 100 feet. The control joint, which is used primarily with masonry, is usually ⅜ to ½ inch wide. Formed as a chase or recess in the wall, it provides a weakened plane through the wall to confine or control cracking. The control joint is spaced at intervals of about 25 feet.

In high-rise curtain-wall construction the major working joints are those between exterior facing panels of the panel-to-mullion joints. These joints become especially critical because they are usually sight-exposed and hence subject to weathering.

In working joints the sealant material has to withstand cyclic stresses over a long period of time without failure. In nonworking joints the sealant functions primarily as a filler and is subject to little or no stress.

Examples of Working Joints

1. Isolation Joints.
2. Control joints in masonry walls.
3. Joints between exterior facing panels.
4. Exterior panel-to-mullion joints.
5. Junction joints, such as that where the sidewalk meets the base of a building.

6. Horizontal joints in terraces, patios, and sidewalks.
7. Lap joints in sheet roofing or siding.

Examples of Nonworking Joints

1. Interior heel beads for glazing.
2. Copings and gravel stops.
3. Reglet joints.
4. Sealing or glazing of small individual window lights.
5. Preformed tape applications in glazing work.

Most working joints can be classified as butt joints or lap joints. Butt joints subject the sealant to alternating tensile and compressive stresses, whereas lap joints subject the scalant primarily to shearing stresses. Figure 3.7a illustrates these two types of movement.

The butt joint is commonly used for exterior panel joints, expansion joints, and control joints in masonry walls. The lap joint is used for exterior panel-to-mullion joints, exterior panel-to-sill joints, and the joints between sheet materials.

Butt joints subject sealants to greater strain. The lap joint permits much greater movement, and also does not subject the sealant to either tension or compression. When a sealant can take ±25% joint movement in a butt joint, the same sealant in a lap joint only takes extension movement and the sealant surface can be displaced 75% of the sealant thickness in either direction—in other words, a ±75% displacement capability. Drawing a sim-

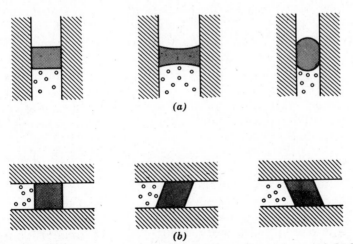

(a)

(b)

FIGURE 3.7a. Deformation of sealant beads in (a) butt joints and (b) lap joints. (Design by Tore Gjelsvik, Norwegian Research Institute.)

ple right triangle will demonstrate the geometry. If the perpendicular distance is 1 inch and then displaced 0.75 inches to either side, the hypotenuse is only 1.25 inches, which indicates a stretch of the sealant of only 25%. Architects could make much use of the lap joint in their design, and have used it to some extent in recent years (see Section 4.2).

3.5.   Calculation of Joint Movement

Although it is generally supposed that most sealants will be applied in moderate weather, in actuality sealants are applied all year around. Only on extreme occasions will work stop when the chill factor has fallen too low or the air temperature in summer has reached too unbearable a high. It is not unusual to see applicators working well below 40°F, and many sealant manufacturers include statements in their specifications that permit application below 40°F with the proviso that steps be taken to assure clean, dry, and frost-free surfaces and that approval of the architect should be obtained.

Even though the air temperature may be high in summer, surface temperature on a dark metal curtain wall facing south or west could be considerably hotter on sunny days. The applied sealants may not necessarily be applied at the extremes in temperature, but can be applied at temperatures close to the extremes. Furthermore, it is not generally undesirable if the sealant is applied during the winter season since the joints will be wider. This may cause the sealant to take a greater compression, but this is more desirable than a greater extension movement after the sealant has been applied during the height of the summer. A sealant applied during the height of the summer will later be subjected to extension beyond its movement capability because the joint was too narrow during sealant installation.

Since butt joints are more common and are generally more critical in terms of the consequence of failure, most computation of joint movement are based on the use of butt joints. It is quite simple to compute a theoretical value for joint movement based on temperature changes. The difficulty arises when one has to determine the temperature gradient affecting the sealant. Figure 3.8 illustrates the temperatures that might be apparent on a dark metal aluminum curtain wall in New York City during the year. If the application temperature and the surface were 70°F, then the greatest temperature differential would be 90°F in either direction. On a 10-foot-long aluminum panel, this movement would be 0.14 inches. For a sealant having a ±25% joint movement, the joint would only have to be 0.56 inches wide. On the other hand, if the sealant were applied on a hot summer day when the surface could very well be 160°F, then the greatest temperature differential affecting the sealant would be 180°F, and the movement affecting the sealant is 0.28 inches in extension only, which means that the

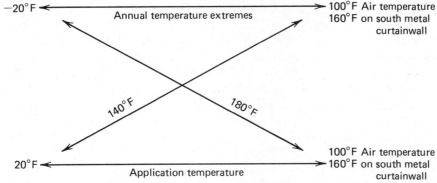

FIGURE 3.8. Determination of maximum possible temperature gradients on a metal curtain wall in New York City.

sealant joint should be 1.12 inches wide *at the time of application.* Since the joint will close 0.14 inches from 70°F to 160°F the designed joint at 70°F should be 1.26 inches wide. Of course, conditions are never quite this extreme, but neither are conditions very near the mean temperature. The example given is based on rather extreme assumptions.

In addition, the surface of a south or west wall of aluminum curtain wall can go through considerable temperature variation even in a single day. Walls facing north or east would not be affected by solar radiation. Walls using stainless steel would expand much less since their coefficient of linear expansion is less than that for aluminum. Concrete, with its lower coefficient and greater mass, is also less affected by heat. Nevertheless, surface temperatures on precast concrete have easily reached 120°F and still give a 140°F temperature differential under the worst conditions. A 15-foot-wide precast panel with the 140°F temperature differential would move 0.15 inches and would require a joint 0.60 inches wide *at the time of sealant installation.* Designing this joint for 70°F would require a width of 0.67 inches if it was expected that the joint might be sealed at the height of summer. Sealant manufacturers generally recommend that the joint width be four times the expected joint movement over the maximum temperature gradient for the area. They do not generally take into account the thermal affect on the surface of the curtain wall due to solar radiation.

The above examples were given to illustrate the problems that can occur, and architects are advised to be realistic about their joint designs and to appreciate the problems involved in application of sealants.

3.6. Slip/Stick Movement in Curtain Walls

The exterior facade of a structure consists of a number of elements, which may consist of several building materials, each with its own coefficient of thermal expansion. Each of these building parts tend to remain stationary until moved by some outside force. In this case, the force is generally expansion or contraction caused by temperature variation. These parts, however, are not subject to uniform movement. As temperature changes, stresses build up within the building part. When these forces become high enough to overcome the inertia of the building part, that part moves. The result is that building joints open and close in a series of short incremental movements, rather than in a uniform cycle.

This slip/stick type of movement is accentuated and aggravated by several factors. The British Building Research Station has calculated the rates of movement of some typical building materials as follows:

Concrete roof coping, 0.020 inches per hour.

Upper brickwork, 0.017 inches per hour.

Lower brickwork, 0.011 inches per hour.

Aluminum mullion, 120 inches per hour.

The extremely fast response of aluminum to temperature change means that the movement of joints of dissimilar materials—such as the panel-to-mullion joints of exterior walls—depends on the rates of movement of both materials as well as the coefficients of expansion.

Another factor influencing slip/stick movement is the temperature variation between the inside and outside of a building wall. The outside surface of the wall may vary from $-20°F$ to $+160°F$; the inside surface of the wall and much of the structural frame to which the wall is attached remains at a relatively constant temperature and humidity. This differential acts to restrain the exterior wall elements, and thereby contributes to nonuniform movement.

There have been a number of independent studies made by various organizations throughout the world that have attempted to determine how joints move. In most cases the general conclusion was that joints move in an unpredictable fashion. In precast concrete wall panels, it was found that some joints behave normally, others are static, and some move twice as much as expected. This occurs when some panels appear to be anchored to adjacent panels so that both move as a unit. Furthermore, in tall precast sections the designed joint width varies with the height of the panel.

The rapid slip/stick movement is not really a problem to most sealants, since the sealant behaves in an elastic fashion with the rapid movement. More damaging is the very slow annual movement when compression set can occur in sealants that have poorer recovery. Burstrom (2) has been able to predict the movement capability of sealants by plotting a continuous curve of stress versus strain on tensile adhesion specimens that have been heat aged for 56 days at 158°F, and then placed in a tensile device at $-5°F$ and separated at the rate of 0.001 mm per minute or ⅛ inch per 24 hours, which would be 25% of the ½-inch-wide specimen. The curve is examined and when the stress reaches a maximum, this becomes the limit of movement capability of the sealant. Burstrom has tested several sealants and found good correlation with his test methods. His study is one of the finest that has been run by any university in the past decade. He has found that polysulfide sealants crack and craze and also show a higher degree of compression set, which lowers his prediction for movement capability as compared to urethane sealants. He did not test silicone sealants since his study was started at a time when silicone sealants were not a factor in the European market. Based on his studies, silicone sealants would rate very high with his test methods.

3.7. Other Factors that Affect Joint Movement

Exterior building joints are assumed to expand and contract primarily as a result of temperature change. However, these joints can also move in longitudinal shear due to the slip of a building panel or differential settlement (Figure 3.9). They may also move in and out in a transverse shearing

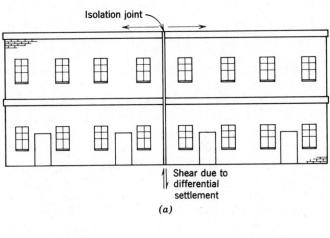

(a)

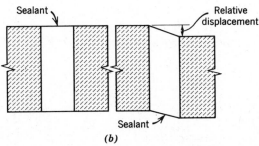

(b)

FIGURE 3.9. (a, b) Longitudinal shear in sealant. Longitudinal shear is a movement of one joint face in a direction parallel to the longitudinal axis of the joint. It could be caused by building settlement or a slight slip of one building panel with respect to the adjacent panel.

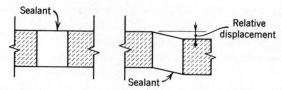

FIGURE 3.10.   Transverse shear in sealant. This type of strain in a sealant can be caused by one building panel moving with respect to the adjacent panel in a transverse direction across the joint.

movement as a result of wind load (Figure 3.10). The sealant between metal building panels may be flexed about its longitudinal axis as the panel edges move in response to temperature changes (the "oilcan effect," shown in Figure 3.11).

The problem is compounded by other effects. The method of attachment of a curtain-wall panel to the structural frame exerts a restraining effect on movement. The color of the panel, moisture absorption, compass orientation, amount of shade on the building, height and mass of the structure, and wind load all play an important part in the movements of building joints. Some excellent beginnings have been made toward incorporating all these effects into a computer program to predict movement (3), but much remains to be done.

Wind load on a building is generally assumed to cause a more-or-less uniform pressure on the windward side of the building and a combination of a uniform suction and internal pressure on the leeward side. However, studies (4 and 5) of air flow around buildings have shown that projecting mullions, other buildings in the vicinity, roof type, and ground conditions all cause turbulent air flow that can stress a sealant in fatigue.

Another factor to be considered in the movement of exterior joints in a curtain wall is the simple statics of building height. A tall building shaft acts much like a huge beam cantilevered out of the ground. When the top of this beam is deflected by wind load, the joints closest to the support (the ground) are naturally subject to the greatest strain (Figure 3.12).

In view of all these factors, together with the scarcity of reliable data on building movement, most architects will rely on the approximate formula and professional judgment to provide reasonable limits for joint movement.

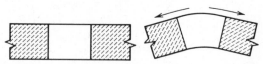

FIGURE 3.11.   Bending (the "oilcan" effect). This is a combination of tension or compression with a displacement of the joint such that the sealant is bent about its longitudinal axis.

Wind
direction

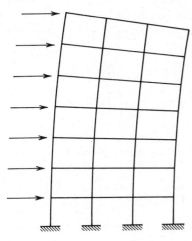

FIGURE 3.12. Wind deflection of a building frame.

3.8. Construction Tolerances

In the erection of a curtain-wall building, the joints are usually well-detailed on the architect's plans. It is quite simple for the detailer in the office to draw the joints to exact scale. At the job site, it is not so easy for even the most conscientious contractor to conform exactly to the drawings. The joint as it is constructed may differ considerably from the contract drawings. The plans may call for ¼-inch-wide joints between concrete panels. The joints as constructed may vary from ¼ to ¾ inches in width. The British Building Research Station has published the following data about tolerances in the length and width of large concrete panels. Note in this table that the total tolerance is not the sum of the individual tolerances. The British Building Research Station has recognized that tolerances may be either plus or minus, and has recommended a total tolerance based on measurements of actual panels.

Type of Production	Site Casting	Normal Factory Casting	Well-Controlled Factory Casting
Tolerance in mold fabrication	±0.12 in.	±0.06 in.	±0.02 in.
Tolerance in slabs cast	±0.32 in.	±0.20 in.	±0.08 in.
Tolerance in erection	±0.24 in.	±0.20 in.	±0.16 in.
Total tolerance	±0.40 in.	±0.28 in.	±0.18 in.

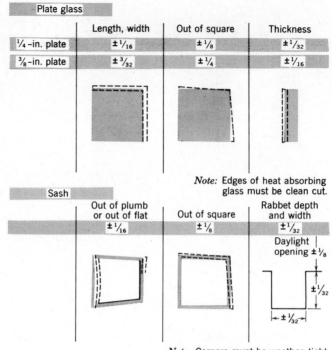

Plate glass	Length, width	Out of square	Thickness
¼-in. plate	±¹⁄₁₆	±⅛	±¹⁄₃₂
⅜-in. plate	±³⁄₃₂	±¼	±¹⁄₁₆

Note: Edges of heat absorbing glass must be clean cut.

Sash	Out of plumb or out of flat	Out of square	Rabbet depth and width
	±¹⁄₁₆	±⅛	±¹⁄₃₂

Daylight opening ±⅛

±¹⁄₃₂

±¹⁄₃₂

Note: Corners must be weather-tight as fabricated.

FIGURE 3.13. Typical dimensional tolerances. (Courtesy of PPG Industries.)

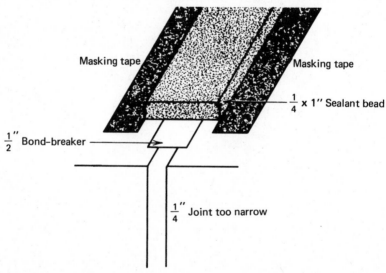

Masking tape

Masking tape

$\frac{1}{4}$ x 1″ Sealant bead

$\frac{1}{2}$″ Bond-breaker

$\frac{1}{4}$″ Joint too narrow

FIGURE 3.14. Bandage sealant design. To be used where expansion joints are too narrow and all sealants would fail due to overextension and compression. Masking tape on edges to be removed either before or after cure of sealant.

Metal curtain-wall panels are factory fabricated under very close control. The total tolerance that might be expected in these panels is less than the tolerance expected in concrete panels. Nevertheless, the metal curtain-wall building may be expected to have joint sizes that vary ¼ inch from those shown on the plans.

Glass panes and metal sash may also vary in size. Figure 3.13 shows typical dimensional tolerances for these units.

Tolerances are important because joint sealant is one of the last materials to be installed in the building. Consequently, the sealant must fill the gaps between the other building components, whether or not these gaps happen to be the proper size. One quarter of an inch does not seem like much of an error in a large building. But consider a joint in a curtain wall; normal joint width is ½ inch. Because of temperature change, the joint is expected to move ⅛ inch in extension and ⅛ inch in compression, a 25% strain on the sealant in either direction. If the joint as constructed is only ¼ inch wide, the sealant will fail, since it is now expected to take ±50% joint movement rather than the ±25% for which it was designed.

In situations where joints have been designed too narrow and any sealant placed in the joint will fail, a bandage type of sealant design has been used to solve this dilemma (see Figure 3.14). The sealant is laid on the surface of the joint. Assuming the joint width is ¼ inch and should be ½ inch, a ½-inch-wide length of masking tape or bond-breaker tape is laid over the ¼-inch-wide joint. Then sealant is applied over the masking tape to a thickness of ¼ inch, and applied in a fillet at least 1 inch wide so that it adheres to the surface for at least ¼ inch beyond the tape on either side. The cured sealant looks like a giant bandage placed over the joint. There is no other solution on metal curtain wall. On concrete or masonry surfaces, it is possible to actually widen out the joint by grinding with diamond-toothed saws. Then bond-breaker tape is laid in and the sealant applied in the normal manner. It may not be necessary to put in a back-up rod since this would entail removing more masonry. Such action is very costly compared to the original installation, but when the leaks occur over the president's desk— and they usually do—the job has to be done at all costs.

Proper supervision at the job site to inspect all materials coming to the site would eliminate such costly blunders. It is not uncommon to see masonry units either oversized or undersized. Undersizing presents no problem to sealant performance, since more sealant is used, but a narrow joint will result in leaking within one season, with all kinds of complex problems arising from water infiltration. Anchor rods have been known to rust and eventually let masonry panels drop to the ground. Such repair becomes a horrendous nightmare.

4

Stresses
and Strains in Sealants

4.1. Strains in Butt Joints

Joint sealing materials play an important part in the overall performance
of a completed building and consequently should be as carefully selected
and designed as any other building component. Many factors will ultimately
affect the performance of a sealant, but the shape and dimensions of the
sealant cross-section are considered of primary importance.

The design of the seal, of course, varies with the type of material being
used. Highly elastic sealants, deformable mastics, tapes, tape sealants, dense
gaskets, and foam gaskets are different materials, and consequently have
different criteria for determining optimum performance.

4.1.1. Highly Elastic Sealants

Silicones, urethanes, and polysulfides can all be manufactured and formulated to give highly elastic and high-recovery sealants. These materials
can also be manufactured and formulated to give sealants that have a much
lower modulus, lower recovery, and higher elongation with a sacrifice of
recovery. It becomes important to distinguish the desirable properties from
the undesirable properties in selecting a sealant, since the low-modulus
sealants might not all be highly elastic.

The highly elastic sealants are prime candidates for exterior, one-stage
sealing operations in working joints because they tend to return to the original shape after the removal of an imposed load. This property of recovery
is most desirable for sealants with a ±25% joint movement capability. Joint
sealants with shallow depth have the best design for reducing strain in the
sealant with movement. Because of sealant design limitations, it is not pos-

sible to give a depth relationship that would hold for all widths. The following four rules would cover most joint designs for the highly elastic sealants:

1. The minimum size joint is ¼ × ¼ inch.
2. For widths from ¼ to ½ inch, the depth should equal the width.
3. For widths from ½ to 1 inch, the depth should be ½ inch.
4. For widths from 1 to 4 inches and up, the depths may be ½ to ¾ inches depending on width and application area as well as sealant.

The minimum of ¼ × ¼ inches is obvious. Joints this small are difficult to seal and anything smaller is almost impossible. The need for ¼-inch depth is to insure adhesion to the side of the joint. From ¼ to ½ inches in width, some manufacturers prefer keeping the depth at ¼ inch, while others require equal width and depth up to ½ inch. At widths over ½ inch, most manufacturers prefer keeping the depth at ½ inch for joints up to 2 inches in width. Some manufacturers may prefer a ⅜-inch depth rather than ½ inch, but this becomes a matter of sealant choice. The sealant depth is controlled by the use of an adequate joint backing, which comes in various compositions and shapes and will be discussed later.

The theoretical shape factor derivation (Figure 4.1 shows a joint with a shape factor of 1) deals only with studies in butt joints and assumes a directly proportional relationship between strain and stress. This assumption is approximately true *only* for the highly elastic and high-recovery sealants, and does not necessarily hold true for the softer, more deformable sealants.

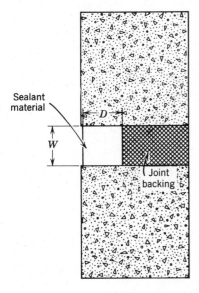

Sealant material

W

D

Joint backing

FIGURE 4.1. Joint with shape factor of one (W/D = 1).

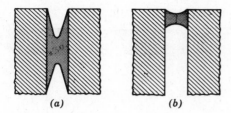

FIGURE 4.2. Comparison of maximum strains. (a) For 2-inch-deep seals. Extension of joint, 100%; sealant strain, 550%. (b) For ½-inch-deep seals. Extension of joint, 100%; sealant strain, 160%. (Courtesy of Egon Tons.)

Quantitatively, the strain in the sealant material may be simply defined as follows:

$$\text{strain} = \frac{\text{deformation of the seal}}{\text{initial joint width}}$$

Figure 4.2 shows two joints with an equal joint width of ½ inch. Both have been extended 25%. The sealant, of course, does not change in volume during extension. Consequently, the deeper seal has to neck down farther when both joints are extended the same amount. Note in Figure 4.2 that the top and bottom surface of the sealant material has extended a great deal more than the corresponding surface in the shallow but proper seal.

When a bead of elastic sealant material is deformed, the top and bottom surface of the sealant neck down into a parabolic shape. Strain can be compared by comparing the lengths of the parabolic curves. Because the top and bottom surface of the sealant undergoes the greatest change in length (strain), the stresses at these joints are naturally maximum. Figure 4.3 shows a qualitative distribution of stress at the interface. Note that the stresses in the substrate at the edges of the sealant bead become quite high. Note also that the direction of the maximum strain (and consequently stress) in the sealant material varies between the deep and shallow seals. In the deep seal, the maximum stress (P) direction has a larger vertical component, and consequently a greater tendency to peel. When this combination of tension and peel stresses exceed the adhesive strength of the sealant material, a failure is initiated. The sealant can fail in adhesion and cohesion, whichever is weaker. On the other hand, if the substrate is a poor-quality concrete, the tensile strength of the sealant could exceed the tensile strength of the

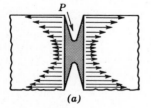
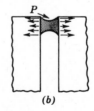

FIGURE 4.3. Comparison of substrate stresses for (a) 2-inch-deep seals and (b) ½-inch-deep seals.

concrete, in which case the concrete would spall along the edge. Either type of failure is undesirable, since the integrity of the joint has been affected.

Sealants can be installed in all climates. The ideal situation is when the sealant is installed in moderate weather, so that the sealant will be subjected to compression as well as tension during an annual temperature cycle. The parabolic deformation of the sealant also holds true when the sealant is in its compression phase. Figure 4.4 shows that the sealant will extrude outside the joint area under load. This may cause an unsightly appearance and—in traffic-bearing joints such as industrial floors—may be extremely critical in terms of both stress and safety hazard. It is advisable in horizontal joints, such as floors and sidewalks, to use a self-leveling material and to keep the level of the sealant slightly below the floor surface.

Since the sealant can be applied in almost any climate, it is possible that sealant applied during the summer in a colder climate would only be subject to extension forces, and that sealant applied in the winter would only be subject to compression forces. This is why the architect must take into consideration the extremes in temperature of the surface (see Figure 3.7) to determine the maximum temperature gradient to use in designing a joint. The architect must design a joint so that the sealant—irrespective of when it is applied—will never be extended more than 25% in either direction. If the sealant is applied in the winter, the sealant might never get any extension during an annual temperature cycle, but might be compressed up to a maximum of 25%. If the sealant were applied in the heat of summer, the reverse would be true: the sealant might never be placed in compression, but would get a maximum of 25% in extension. If by some stroke of luck the sealant could be applied at the mean temperature, the sealant could go through a half season of increasing compression to a maximum of 25% in the midsummer, and then begin reversing the cycle until in midwinter it would be extended a maximum of 25%.

An improvement in the rectangular cross-section can be achieved by using a curved back-up material and tooling the top surface as shown in Figure 4.5. This tooling of the joint must be done with some care, however, because if the sealant bead is too thin at its center, the sealant under compression will buckle like a slender column. This will cause high peel stresses at the lower corners (Figure 4.6).

In addition to having the proper shape factor, the elastic sealants must be bonded to only two sides of the joint in order to perform properly. The bottom surface of the sealant must be free to deform. Figure 4.7 shows

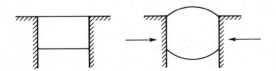

FIGURE 4.4. Sealant bulge under compression.

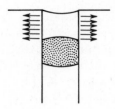

FIGURE 4.5.   Uniform substrate stress in tooled joint.

that if the bottom of the joint is bonded, the sealant must rupture in order to deform. Figure 4.8 shows that the same basic principle of permitting the bottom of the sealant bead to deform also applies to corner joints.

These high-recovery sealants have the advantage of returning to the original shape after the removal of an imposed load. This means, then, that these materials remain in a stressed state as long as they are deformed. As a generalization, it is true with rubbery sealants such as silicone that the higher the recovery, the lower the tear resistance. Consequently, once a failure is initiated in the material, the failure is likely to propagate rapidly, thereby causing extensive damage.

4.1.2.   Deformable Sealants

The deformable sealants cover a wide range of materials: polysulfides, polymercaptans, butyls, solvent-based acrylics, solvent-based chorosulfonated polyethylene, and "latex" caulks of various composition. The chief characteristic of these sealants as a group is that they show some degree of instantaneous elasticity or "rubberiness" under short-term loads, but will creep under long-term loading. Within the group, the sealants exhibit these two properties in varying degrees. Polysulfides, for example, may be highly cross-linked to show a large percentage of instantaneous elasticity (recovery) and only a small amount of cold flow. On the other hand, the solvent-based acrylics will show very little recovery.

The deformable sealants are generally installed according to the general shape factors laid down for the highly elastic sealants. The one limitation is that the maximum width is usually never more than ¾ inches. This is apparent since the sealant could begin to show some permanent change with several annual temperature cycles. The deformable or low recovery sealants show a great deal of stress relaxation, and when held in a deformed

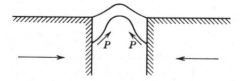

FIGURE 4.6.   Peel stresses in the lower corners of an overtooled joint.

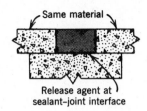

Same material

Release agent at
sealant–joint interface

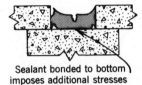

Sealant bonded to bottom
imposes additional stresses

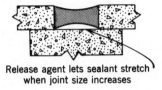

Release agent lets sealant stretch
when joint size increases

FIGURE 4.7. Release agent at bottom of joint eliminates three-sided adhesion. (Courtesy of Machine Design.)

state, the materials will relax into equilibrium so that a new shape is formed in this unstressed state.

The other limitation for this class of sealants is that they are now limited to areas of application in which movements do not exceed $\pm12.5\%$. The various manufacturers will vary in their maximum movement recommendations from $\pm7.5\%$ to 12.5%, but they recognize the limitations for this class.

By limiting the class of sealant to joints with less movement, the sealant does not become very distorted with repeated temperature cycles. Another property with most sealants in this class is that with the loss of solvent, some sealants become quite tough and can get to Shore A hardness of 50 to 60, which is not conducive to large movement. This group is used in limited-movement joints because it has adhesion to most surfaces without the need

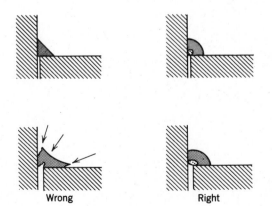

Wrong

Right

FIGURE 4.8. Design of corner joints. (By Tore Gjelsvik, Norwegian Building Research Institute.)

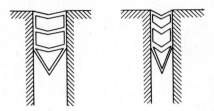

FIGURE 4.9.   Preformed seals.

for primer. The highly elastic sealants may require primer in various areas; urethanes as a group are not recommended for glass and require a primer for metal: silicones may require a primer for concrete; and polysulfides are not recommended for glazing. The high solvent content in this group permits the deformable sealants to "wet" most surfaces and no primer is required. These sealants are also used on wood, baked finishes, and even moist surfaces with good adhesion. The wide versatility of the solvent-based sealants accounts for their popularity, and when properly used within their joint-movement capability, they do an excellent job. Some manufacturers claim a "self-healing" property for their sealant—but proper usage would not result in excessive extension.

The calculation of stresses on the deformable sealants has been accomplished (6), but the results are of academic value only. What is important is the amount of shape distortion to which these materials are subject. It is a paradox of joint sealing that these materials, which can generally take more ultimate elongation than the highly elastic sealants, should be used

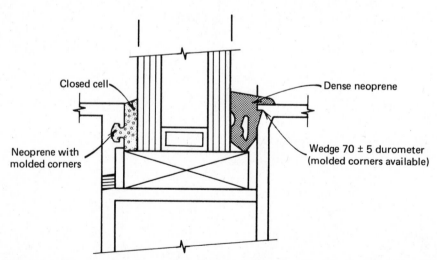

FIGURE 4.10.   Compression glazing system. Closed cell gasket to exterior with Dense neoprene gasket to interior.

preferably in joints with less movement. Stress is of little consequence since it relieves itself so rapidly. However, if these deformable sealants are subject to excessively large deformations, they tend to wrinkle at the surface and cause an unsightly appearance.

4.1.3.   Preformed Shapes

The preformed seals, or gaskets, are generally formed from high-recovery elastomers such as neoprene or EPDM as either high-hardness, dense-rubber extrusions, or closed-cell sponge gaskets of various densities. Because of the principle under which they operate, they depend on high recovery in order to achieve optimum performance. In some joints these seals are

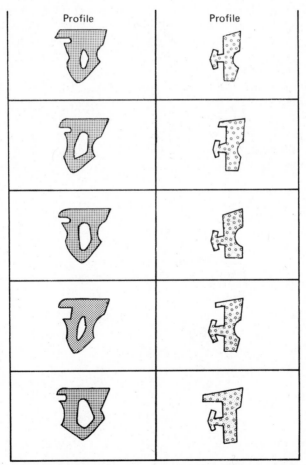

FIGURE 4.11.   Examples of various closed-cell and dense gaskets. (Courtesy Tremco, Inc.)

compressed and then installed into the joint in a compressed state. As they attempt to return to normal shape they exert a force against the joint wall, thus forming an effective seal (Figure 4.9).

In the area of glazing, a new technique based on dry/dry glazing uses two types of preformed shapes. Figure 4.10 shows a sponge gasket to the exterior, and a dense gasket to the interior. The dense gasket is in a wedge shape that creates a pressure against the glass and the sponge gasket—and the high recovery maintains a seal on both sides of the glass. The shapes of the gaskets are numerous on the metal design. Some metal designs permit interlocking of the gaskets to prevent their being worked out of the opening with glass and metal movement. Figure 4.11 shows the wide variety of glazing gaskets available today. Brown (7) has shown that a typical neoprene closed cell foam will require a pressure of 15 psi to compress the seal 25%. This value will decay to 5 to 6 psi after several years of service. This level of sealing pressure is more than sufficient to keep the joint well sealed.

4.1.4. Tapes

Tapes are made of either a cured (high-recovery) or uncured (low-recovery) material. There is no stress in an uncured tape that is desired for good and permanent wetting. Resistance to flow into very thin films in most cases becomes the resisting force whereby the uncured tapes become good void fillers in metal building overlapping seams. Where the force would become a major factor, many tapes are designed to include a cured rod of butyl rubber, which limits the movement and results in both a mastic seal and a

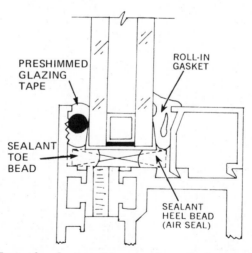

FIGURE 4.12. Example of a pre-shim tape in a glazing application to prevent extrusion of mastic sealant with high wind load. (Courtesy Tremco, Inc.)

semirigid seal. Since the mastic material is generally a butyl/polyisobutylene material, butyl rubber is generally used in the rod since it is compatible, as compared to neoprene or EPDM, which may contain incompatible ingredients and may also be more expensive. Such materials are the pre-shimmed tapes as shown in Figure 4.12.

4.2. Strains in Lap Joints

Stresses and strains in lap joints have received very little attention. This is probably due to the difficulty in designing lap joints in structures and particularly in curtain-wall panels. It might be possible to design lap joints in the vertical joints and butt joints in the horizontal joints. The lap joint has much greater movement capability with much less strain on the sealant. Whichever direction the bonding surfaces move with respect to one another, the sealant is only subject to shear. The proper shape factor for use in lap joints has been largely a matter of common sense and judgement. Koppes (8) recommends that the thickness of the sealant in the joint be at least equal to the amount of the expected total joint movement.

Simple geometry would illustrate that as the thickness of the sealant increased with respect to the expected joint movement, the stress would lessen. The only limitation would again be sealant cross-section and design. Common sense would indicate that the minimum size of the sealant cross-section should be ½ × ½ inches. This joint could easily take ±¼-inch displacement or a 50% displacement in either direction, with the sealant only being extended 12%. This is illustrated in Figure 4.13a, which has the hypotenuse extended only 12%, giving a 50% displacement in the lap joint. The hypotenuse is the sealant edge after displacement. If one assumes that the sealant is capable of being extended 25% along the hypotenuse, which is a very obvious assumption with a ±25%-movement-capability sealant, then Figure 4.13b shows that the ½ × ½-inch joint can be displaced 0.378 inches or 75%. If the joint width is widened to give a ¾ × ½-inch joint and assuming only 12% extension of the sealant along the hypotenuse, this gives a 75% displacement as shown in Figure 4.13c. Finally, assuming a 25% extension of the sealant along the hypotenuse, the displacement becomes 150% in either direction, as shown in Figure 4.13d. In all sealant installations, the architect must consider the maximum temperature gradient. In this design, the architect has considerable leeway because of the much greater movement capability using lap joints.

Lap-joint movement occurs on the glazing sealants and tapes placed between the glass and the stop. This movement can roll glazing tapes out of the joint if movement is too large. This has been solved in part by designing complex shapes in the metal and rabbet or by using nubs or keys for holding tapes and gaskets. In some cases, a cap bead of silicone has been used to keep the tapes in position with movement.

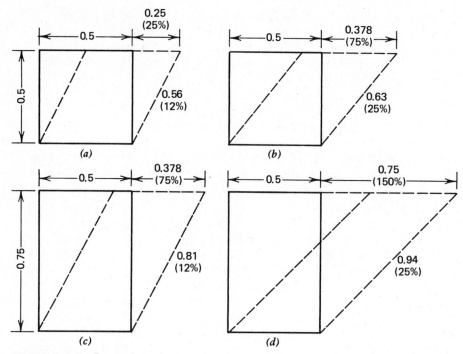

FIGURE 4.13.   Lap shear displacements. (a) and (b) are ½ × ½-inch joints showing 25% displacement with 12% extension of sealant, and 75% displacement with 25% extension of sealant. (c) and (d) are ¾ × ½-inch joints showing 75% displacement with 12% extension of sealant, and 150% displacement with 25% extension of sealant.

4.3.   Strain in Shear Movements

Sealant placed in butt joints, although mainly subjected to extension and compression movement, can occasionally be subjected to shear movement that can be devastating depending on the location of the sealant. In curtain-wall panels, there is practically no shear movement unless the various panels are anchored differently. This could occur if two adjacent panels were anchored such that one was anchored at the top and the other at the bottom. Then as the panels expanded in opposite directions, the sealant would be placed in shear. Theoretically, shear movement calculation can be treated the same way as for butt joints.

A problem area in which failure in shear has occurred is in parking decks where the expansion joints may not be properly placed. Specifically, there may be several long joints with side joints running into the long joints but not through them. At these junctures, the sealant is torn in the center of the joint for respectable distances. The tear can be so sharp as to appear to be cut with a razor blade. Shear stress also occurs at all intersecting joints.

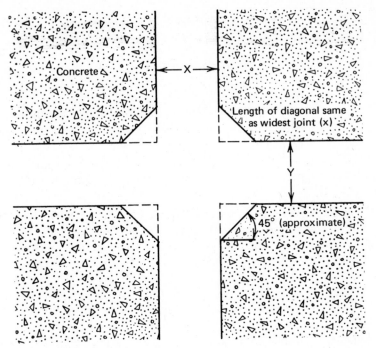

FIGURE 4.14.   Treatment at intersecting joints.

This can be resolved by cutting out the corners and putting more sealant at the intersections (Figure 4.14). The problem in plaza joints can be alleviated by making all joints continuous and putting in more joints. This then only leaves the joints that terminate against a building. The joint next to the building should be wider than normal to take into consideration possible shear movement at these termination junctures.

4.4. Types of Failures

The types of failures that can occur in a sealant depend on a number of factors and the type of sealant being used, as described in the next two sections.

4.4.1. Fluid-Applied Sealants

Adhesive failure, which is the most common type of sealant failure, is a loss of bond between the sealant material and its substrate, as shown in Figure 4.15. Adhesion failure can occur in one of several ways:

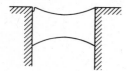

FIGURE 4.15.   Adhesion failure.

1.   The sealant was improperly formulated, and therefore did not adhere to the substrate. With the multitude of substrates available, this is not unusual.
2.   The sealant became tough and more limited in its movement capability, and eventually failed because it could no longer take the required movement.
3.   The joint was made too narrow, so that the sealant never had a chance. Although at first glance this appears to be a sealant failure, inspection of the joint, panel sizes, and temperature gradients indicate a design failure. In essence, a sealant cannot hold a building together.

Cohesive failure is a failure within the body of the sealant material. This failure frequently begins with a small nick or puncture of the material (Figure 4.16). Because the sealant did not fail in adhesion, at least this property was properly formulated into the sealant. The cohesive failure could occur for reasons 2 and 3 above. Also, with bulk sealants, the sealant is soft, and with repeated movement—usually beyond its movement capability—the sealant parts in the center.

Spalling failure is not always a sealant failure, but its effects are just as destructive. If the cohesive strength of the sealant is greater than the cohesive strength of the substrate, this type of failure will occur. Spalling failure is frequently seen with stiff, high-recovery sealants when used with low-strength concrete panels (Figure 4.17). It is possible for a sealant that cures to a high modulus to spall good quality concrete, since the tensile strength of good concrete is approximately 300 psi and some sealants can exceed this value. Such sealants would be improperly formulated and if placed on metal surfaces would fail for reason 2 above. Spalling is one of the most frequently encountered failures in highway pavement joints, but a number of factors are involved (for more information see Chapter 19).

Intrusion failure (Figure 4.18) is seen in highway pavement joints and other traffic-bearing areas. This failure occurs as a three-step process: (1) the sealant extends, necks down, and then fills with dirt; (2) during the

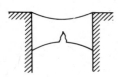

FIGURE 4.16.   Cohesion failure.

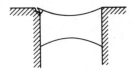

FIGURE 4.17. Spalling failure.

succeeding compression cycle, this pocket of dirt attempts to close; and (3) this closing action generally abrades the surface of the seal so that it fails during a subsequent tension cycle.

Two additional types of failure are common, both of which are peculiar to the deformable sealants. One type occurs when compression is the first strain to be applied. The tensile failure occurs as follows. The material is extended and held in a necked-down shape (Figure 4.19). The sealant relaxes into equilibrium in this new shape. When the joint closes, the sealant acts like a slender column and buckles. In addition to exposing the strained sealant to further weathering, this may cause high peel stresses at the lower corners. The compression failure is shown also in Fig. 4.19. The sealant is compressed and relaxes into equilibrium in this new shape. When the joint opens, the sealant does not return to its former rectangular shape; instead, it yields at its minimum cross-section, which is immediately adjacent to the joint interface. This highly concentrated strain may lead to immediate failure.

Remember too that the stress and strain computations that have been made concerning bulk sealants all assume that the sealant is fully cured before it goes to work. This assumption is far from true. The building, bridge, or pavement is not so obliging as to wait until the sealant cures before it begins to expand and contract. The structure is constantly moving, and the sealant must be expected to function as soon as it is placed in the joint. The effect of this joint movement on the uncured sealant has never been investigated fully, and is not considered in any of the standard specifications.

However, with multipart sealants a large portion of the cure has been affected within 24 hours—and within any diurnal temperature cycle, total joint movement is not anywhere as extreme as the annual cycle. Most two-part polysulfide sealants will have less than 2 hours of work life when temperatures are above 80°F. Multipart urethane sealants will also have a fast

FIGURE 4.18. Intrusion failure.

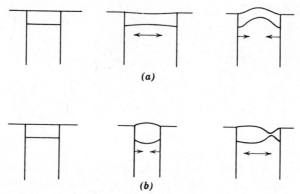

(a)

(b)

FIGURE 4.19.   Change in sealant shape due to flow. (a) Viscous tension-compression effect. (b) Viscous compression-tension effect.

cure when the ambient temperature is above normal. During the cooler climates, cure will be slower, but movement will also be lower.

With one-part sealants, whether moisture-cure or solvent-release type, a skin forms that tends to keep the sealant intact until complete cure has taken place. With one-part silicone sealants, skin formation can occur within a matter of minutes. With moisture-cure urethanes the skin can occur within a matter of several hours and the sealant is tack-free in 8 to 12 hours. With solvent release sealants, tack-free time may be 1 to 3 days under normal circumstances. Overall, cure rate has been no problem even with small joint movement, since generally sealants attain some degree of integrity because of the skin formation. This is based on the assumption that joint size is proper.

4.4.2.   *Preformed Seals*

The most common failure of the preformed seals is a loss of interface pressure between the seal and the substrate. The usual cause of this failure is simply poor design, which results in the wrong size seal being placed in the joint. In order to perform properly, the seal must continue to exert pressure against the interface when the joint is at its widest. On the other hand, when the seal is compressed the voids may be closed, but the seal should not be compressed so far that the rubber itself is stressed in compression.

The preformed seals are generally fabricated of high-quality elastomers; hence failures within the seal itself are rare. Some instances have been noted of extrusions cracking after weather exposure, but these have generally been traced to poor rubber compounding. There are several excellent ASTM standards for closed-cell foam gaskets and dense gaskets—specifically, C-864, entitled "Dense Elastomeric Compression Seal Gaskets, Setting Blocks, and Spacers," and C-509, "Cellular Elastomeric Preformed Gasket and Sealing Material." These standards and others have set the standard for industry.

The preformed seal, however, depends on straight uniform joint walls in order to function. Consequently, when the extrusions are used in conjunction with metal curtain-wall panels, any warping or nonuniform distortion of the panel causes the seal to lose its effectiveness. When they are used with concrete panels or slabs, as in highways, industrial floors, or terraces, any spalling of the concrete at the joint causes the seal to be ineffective.

The foams that are used as seals are generally of closed-cell urethane or neoprene. These materials have the advantage of being less rigid than the extrusions, so that straight joint walls are not so critical. In addition, there is little or no lateral spread to these materials when compressed and therefore they do not extrude out of the joint. However, the foams are subject to a slight amount of flow and stress relaxation, which may lead to some loss of seal effectiveness over a long period of time.

4.5. Stress Determination by Experimental Methods

The actual determination of the stress distribution within the body of a sealant material is difficult to establish analytically. When dealing with crystalline materials that undergo small deformations, it is relatively simple to use strain gauges and thus determine stress experimentally from the modulus relationship. With the rubbery materials, however, the deformations are large and the stress–strain relationship is nonlinear, so that analytical methods become complicated and tedious.

Visual methods of strain determination can be used very effectively with sealant materials. One simple method of determining a qualitative strain distribution requires only a testing machine and a felt-tip marking pencil.

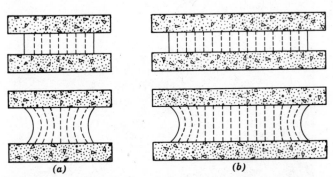

FIGURE 4.20. (a) 2-inch-long and (b) 4-inch-long specimens, both extended 100%. Dashed lines marked on the specimen clearly indicate the directions of the strains and stresses in the sealant.

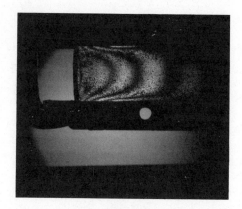

FIGURE 4.21. Photoelastic stress patterns in a sealant specimen under compressive load.

FIGURE 4.22. Photoelastic stress pattern in a preformed joint seal.

Lines are drawn on the specimen at equal intervals. When the specimen is extended, the directions of the strains are immediately apparent. Figure 4.20 shows the differences in the strain directions in two sealant specimens. Although the use of longer specimens might be better from a theoretical standpoint, preparation of laboratory test samples and designing an appropriate test apparatus favors the smaller sample. Proper interpretation of test results is far more significant.

Quantitatively, the method of photoelasticity offers the best solution for the stress in a sealant. With this experimental method, it is possible to literally look inside the specimen and see the stress distribution. Certain translucent materials, such as urethanes and polysulfide epoxy combinations are artificially birefringent materials. This means that these materials will show a pattern of stress lines when loaded and placed in a polariscope. Thus, specimens in proper size and shape can be cast of this translucent material, and the stresses can be determined experimentally. Correlation between the translucent models and the opaque sealant material is then established by plotting stress–strain curves for the two materials. Figure 4.21 shows the stress pattern in a loaded specimen. Cook (6) has successfully used this method to verify Ton's original work on shape factors; to determine the stress-relaxation rate of polysulfide sealants; to determine the stresses in preformed highway seals; and to study the length of sealant specimens. Research is currently under way to determine the optimum shape factor for shear stresses in lap joints and also to determine the maximum stresses in structural glazing gaskets. Figure 4.22 shows the stress pattern in a preformed seal. Because of the great variety of shapes in use, the determination of stress is exceedingly complex. The use of photoelasticity is also very useful here.

4.6. Fatigue

Fatigue in a construction material is essentially failure caused by a large number of repetitions of stress. The failure may be the result of crystalline alignment within the metal, with the creation of unequal stress points and eventual rupture along the weakest plane. In sealants, fatigue involves not only stress and strain, but compression set, hardness increase, and surface degradation. Burstrom (2) has done some excellent work showing cohesive failure or fatigue of several classes of sealants. His studies show that during the summer season, the sealant takes greater increments of compression, accompanied by accumulated increments of stress relaxation that varies with different sealants. During the winter, the samples receive greater increments of elongation after maximum compression set has occurred at the height of the summer season. The sealant may either begin to neck down or show some adhesion loss due to the new configuration. Failure can occur after several years if the sealant is prone to compression set, increases in hardness, and has surface degradation. Surface cracks from ultraviolet light can become focal points for further cracking. These results indicate the need for sealants that do not harden, or toughen with time, have low modulus, high recovery, and are not affected by ultraviolet light or ozone.

Karpati (9, 10) in several studies has done some excellent work on sealants on stress cycling expansion racks. In a 3-year study with silicone sealants, samples with greater than ±22% movement showed a greater incidence of failure. These studies were run on a one-part high-modulus silicone sealant. Most failure was in adhesion starting at the corners of the tensile adhesion specimen. Her conclusions were that stress-cycling movement was the predominant factor that caused failure, and that outdoor weathering alone was negligent. Karpati observed that sealant installed in the spring did not show failure after 3 years of exposure because the adhesive bond was firmly established before exposure to string cycling. The statistical evaluation was an excellent study, and perhaps repetition using a lower-modulus silicone sealant would result in better movement capability.

5

Properties of Sealant Materials

Sealants are tested for a number of properties—some of these tests are useful in describing the desirable properties of sealants to the consumer. Other properties are studied with the intent of improving long-term sealant performance. Sealant manufacturers may use descriptions of properties that architects can understand but that are not meaningful in predicting performance. This is unfortunate, since many of the more meaningful performance data may be left out because the information might be misinterpreted by the architect or user. Emphasis will be placed on certain desirable properties that can be incorporated into performance specifications.

There is no such thing as an ideal sealant, since no one sealant has all of the desirable properties for the various applications involved. However, it must be realized that ease of application—such as one-part versus two-part—plays a major role in sealant selection, particularly in the home-repair market. Price, which should not be a major concern, nevertheless does influence buying at many levels of purchasing.

A partial list of some of the chemical and physical tests used to evaluate sealants is shown as follows:

Movement capability.
Hardness.
Ultimate tensile strength and elongation.
Tensile adhesion.
Solids content.
Adhesion-in-peel.
Resistance to water immersion.

Resistance to heat aging.

Resistance to weathering.

Compression set resistance.

Tear resistance.

Modulus of elasticity.

Abrasion resistance.

Solvent and chemical resistance.

Water resistance.

Toxicity.

Electrical properties.

Cyclic tension and compression.

All of the tests are valid and generally defined in terms of either some ASTM test requirement or the requirements of one of the reference standards. Out of this maze of properties and tests, each defined in its particular unit, it is very difficult for even experienced architects or engineers to know what they are buying.

The more important properties and methods of test will be defined in order to facilitate the selection of the proper sealant for specific applications. The limitations will be pointed out. Quantitative values for the various tests as they apply to specific sealants will be outlined in the chapters covering the respective sealants.

5.1. Movement Capability

Considerable effort has been made to emphasize the movement capability of a sealant in a joint. The value is always stated as a ± percent value that indicates the amount of movement the sealant can take either in extension or compression *from its original installation.* It is important that the architect recognize that the sealant can be applied in any season. If a sealant is applied in summer, it may be subjected mainly to extension forces. Sealant applied in winter will be subjected mainly to compressive forces. Only in the spring or fall will the sealant be likely to get equal extension and compression. The three generally recognized categories are ±5%, ±12.5%, and ±25%. Work is being done on sealants that might have greater movement capability. In every category, however, architects should beware, since the maximum must not be surpassed, and all tolerances must be considered in joint sealant design.

Work has started in ASTM Committee C-24 for a ±50% movement class for low-modulus silicone sealants. There is also some indication that this movement can be met with modified urethane sealants.

FIGURE 5.1. Shore A durometer.

5.2. Hardness

The hardness of a sealant is not in itself an indication of how the material will perform in a building joint. However, the hardness is a measure of the resistance to penetration, and is used as a quick measure of the state of cure and the modulus of elasticity. Hardness is measured by a Shore Durometer (Figures 5.1 and 5.2). The scale of 0 to 100 on the Shore A scale is strictly relative. For example, soft sealants that have high movement capability would fall in the 15 to 25 Shore A range. After aging, these sealants should not get above 45. Thus hardness is a relative scale of toughness and aging. Architects should specify a maximum hardness of 45 after 6 weeks' heat aging at 158°F. Solvent-release sealants will show a low initial hardness

FIGURE 5.2. A pencil Shore A durometer.

range of 12 to 25; but after several years of additional aging, the hardness should not exceed 50, since the sealant will lose its ability to take maximum movement. Reading the instrument is very important and the data should state either instantaneous hardness or hardness after 1 or 2 seconds. It is very important that the architect be aware of this point. ASTM C-661 requires an instantaneous reading, while ASTM D-2240 requires the reading within 1 second (see paragraph 7.2 of D-2240). There could be a 5-point difference because of this time interval.

To better illustrate hardness, an automobile tire has a hardness of about 70. An inner tube has a hardness of about 40 and a pink rubber eraser of about 15. The hardness of sealants varies with temperature and there could be an increase of 10 to 20 points with some sealants at $-20°F$. This increase should be kept in mind when selecting sealants for cold climates.

5.3.   Ultimate Tensile Strength and Elongation

Many sealant manufacturers' data sheets will show ultimate elongation, with values ranging from 500% to 1000%. However, this property is derived on a thin dumbbell specimen 0.08 inches thick and is really used for testing rubber samples for tires, hose, molded goods, and so on; and performed using ASTM D-412. This test method is valid for rubber goods but has no place for joint sealants where the maximum movement is in the order of ±25%. These high values should not be reported.

The same holds true for ultimate tensile strength. The high values of 200 to 500 psi listed for ultimate tensile strength are also misleading and useless when used to describe sealant properties. Movement capability is the major concern, and this should be measured on a tensile adhesion specimen.

5.4.   Tensile Adhesion

The adhesive strength of a sealant and a measure of the extensibility of sealant in a joint configuration may be measured by a tensile adhesion (bond-extension) test. ASTM C-920 contains a bond-cohesion test for initial adhesive strength, and cycling after initial compression and water immersion. Another test reports cycling after ultraviolet exposure. The text assemblies are shown in Figure 5.3. Since the standard test specimen for a joint sealant is ½ × ½ × 2 inches, the values for 12.5%, 25%, 50%, and 100% might be listed for tensile adhesion. Since the sealant has a 1 square inch cross-section, there is no need to convert to psi. The obtained values are directly given in psi. Low-modulus and hardness sealants might list values in the 5 to 10 psi range for extensions up to 100%; 15 to 30 psi for medium-modulus

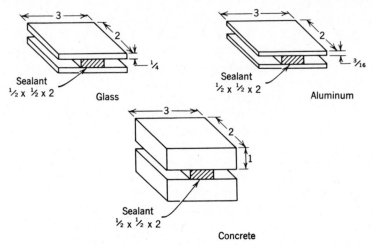

Sealant
½ x ½ x 2   Glass

Sealant
½ x ½ x 2   Aluminum

Sealant
½ x ½ x 2

Concrete

FIGURE 5.3. Typical bond specimens.

and hardness sealants; and 25 to 50 for the high-modulus and hardness sealants. Architects should stress that the tensile adhesion specimen in ASTM C-920 be performed on specimens that have received initial cure, and on specimens that have been heat aged 3 weeks at 158°F.

The criterion for determining failure as spelled out in the C-920 standard is vague and difficult for the potential user to evaluate. Failure is defined as an adhesive or cohesive separation exceeding 1/10 inch in depth. Potential users know only that the material "passed" this requirement. They have no way of knowing how the material actually survived the test. For example, the material may develop a slight tear or loss of adhesion and still pass the test. However, most rubbery materials if notched or punctured will fail completely within the next one or two extension cycles. Performing the additional tests after heat aging will still be a major step in insuring quality. This requirement is not included in any standard at the present time.

5.5. *Solids Content*

Solids content is reported for most sealants, and most sealants meet present specifications. The present values are too high even though they might seem low. The volatile components in a sealant are solvent and volatile plasticizer—which have specific gravities less than 1.0. Most sealants will weigh from 13 to 16 pounds per gallon or have close to a specific gravity of 2.0. In round figures, a 10% weight loss will result in a 20% volume shrinkage. This large volume shrinkage will in most cases be associated with an increase in hardness, excessive stiffening, and an increase in toughness that could result in a loss of some of the movement capability of the sealant. Many sealants require a small amount of solvent to help in the extrusion properties of the sealant, and the non-sag property when the solvent evaporates. The solvent-release sealants depend on solvent to dissolve the polymer to bring it to an extrudable form. Then upon release of the solvent, the polymer returns to its tough state. However, the permitted use of high solvent content also permits the sealant manufacturer to put in more filler and cheapen the sealant.

The sealant would be of much better quality if the volatiles were kept at a minimum. Sealants having ±25% movement capability should have a maximum of 6% weight loss after 6 weeks aging at 158°F. Sealants having ±12.5% movement capability should have a maximum of 12% weight loss under the same conditions. Decreasing the volatiles would lower the volume shrinkage.

5.6. Adhesion in Peel

The peel adhesion test when used in addition to the tensile adhesion test gives a good indication of how a sealant might perform in service. Lap joints that undergo a shear-type movement exert a mild peeling action on the bond surface of the sealant. None of the standard specifications include a shear test. Although the peel test is not strictly a shear test, it does give some indication of the adhesive strength of a sealant in shear.

Specimens for the test are usually 1 inch wide and about 6 inches long. The specimens are formed on a substrate—either aluminum, glass, or masonry—and a strip of cotton fabric. Some specimens use a wire mesh for long immersion since the cloth will rot. The cloth is rolled into and impregnated with the sealant. After the sealant has cured, the cloth strip is folded back 180 degrees and the cloth and sealant attempted to peel back at a rate of separation of 2 inches per minute. The load is reported in lbs per inch of width.

Because this is a destructive test, the results cover a wide range, and duplication from laboratory to laboratory can vary considerably. However, the variability could be eliminated if the test were run as a nondestructive test by suspending a dead weight to the cloth and noting whether adhesion or cohesion failure occurs with any specific weight. This would eliminate the wide spread of the data.

The real value of the test is that it is an easy way to determine the adhesive character of the sealant. Tests can be made on any substrate, either in the laboratory or on the job site. This eliminates the need to run the tensile adhesion assemblies on all surfaces. With the multitude of building surfaces available today, this is the one quick and easy test that can be run to determine adhesion. Minimum values of 5 lbs per inch of width are used in ASTM C-920, which is considered satisfactory.

5.7. Resistance to Water Immersion

Water immersion of most test specimens before test is now a part of most standard specifications. This is valid since water or moisture are encountered in sealant applications at some time. With masonry, the water can reach the interface and ultimately cause adhesion failure. While many immersion tests are based on 1 week's immersion, it is interesting to note a major difference in test results between 1- and 2-week water immersions. Consequently, 2 weeks is the recommended time for all water immersion tests. Many European standards require immersion in alkaline water, since most of the sealant applications are on masonry. With some sealants, there has been a marked deterioration of the sealant and adhesion when applied to masonry that has been wet for periods of time. The conclusion is drawn that water extracts some of the alkalinity from the masonry, which makes the water more damaging. With the increased use of masonry as a building material in the United States, this modification may have some real value in domestic standards.

5.8. Resistance to Heat Aging

The effect of high solvent loss has been discussed. High surface temperatures of 158°F and above have been recorded on many metal curtain walls facing south and west during the summer months. Sealants on these exposures will receive some heat aging, and the total exposure could amount to 6 weeks at 158°F after 2 years. For this reason, many test specimens should receive additional exposure and should receive 6 weeks at 158°F aging prior to test to ascertain that the sealant will exhibit longevity in performance. Some sealants continue to polymerize and cross-link which results in tougher sealants. The recommended exposure is merely the application of an accumulated environment in a shorter period of time. This is not an exaggerated test environment. Specification writers would be wise to increase exposure time in all of their specifications.

5.9. Resistance to Weathering

Ozone and ultraviolet (UV) radiation are real factors in environments throughout the world. Many major cities now exhibit 50 pphm (parts per hundred million) of ozone in their atmosphere, which is the level set in standard ozone environmental test equipment. Ozone at this level will cause surface crazing and cracking after reasonable exposure. It is a known fact that UV light will cause some sealants to lose adhesion to glass in glazing applications. The UV is reflected to the surface of the sealant against the glass. Polysulfides and some urethanes have exhibited weakness to these

FIGURE 5.4.   An Atlas Weatherometer.

environments. UV is stronger as one approaches the equator. Resistance to surface degradation after several thousand hours in a standard weatherometer (Figure 5.4) is indicative of good performance. Burstrom found that in general the effect of UV degradation was a surface phenomenon, but did not discount the possibility that surface crazing with joint movement could result in increasing the depth of surface cracks. For these reasons, architects should be aware of environmental test in weatherometers and the number of hours of exposure. Also, some adhesion tests are run on glass—with the sealant immersed—while the upper surface of the glass is dry and exposed to a UV light source. This wet UV adhesion in peel test to glass is a standard part of many sealant standards when the sealant might be considered in a glazing application.

5.10. Compression Set Resistance

This is an extremely easy test to run. Two parallel plates hold a piece of cured sealant under any prescribed set of conditions. The Hockman test cycles in ASTM C-920 include a compression cycle in part of the test exposure. Data sheets from some manufacturers include compression set data. The omission of this data from some data sheets is not an error, but merely an omission because the sealant may have poor compression set resistance. Although structural silicone sealants have high recovery, which is very desirable, the low-modulus silicone sealants may or may not require it. Polysulfide sealants show poor recovery—yet the better sealants meet ASTM C-920. Urethane sealants can exhibit good recovery and meet ASTM C-920. Exactly how much recovery or compression set resistance is required is not known, but better performance is expected if recovery is good.

Tests by Karpati (10) on high-recovery silicone sealants showed tear when samples were continually tested to over ±22% movement on an outdoor test rack. Below this value, the sealants performed satisfactorily. No tests have been reported on the lower modulus sealants and probably higher compression set silicone sealants. However, some amount of elastomeric memory is desired for performance in high-movement joints.

5.11. Tear Resistance

Tear resistance is a relatively important property, but as a pure test is not as important as other requirements. Generally, if the modulus of the sealant is lowered, the sealant is less apt to tear. The tests are run on died out specimens from thin rubber sheets, and the test samples are usually nicked. The test samples may have a "V" shape, and the samples are pulled apart on a tensile tester and the value measured when tearing starts. The results are reported in lbs per inch of sample thickness. Tear resistance is an important test for tire and hose compounds, but only of relative importance in sealant compounds. Sealants that harden after aging would be more subject to tearing since tearing is a common failure with higher-modulus compounds.

5.12. Modulus of Elasticity

The modulus of elasticity of a sealant material is one of the more important properties of the material to the laboratory technician, but at the present time is of little value to the specifier in the selection of a sealant.

The modulus of elasticity of any material, or the ratio of stress to strain, is determined experimentally as the slope of the stress–strain curve at some point. Most materials, such as metals, have a stress–strain curve that includes

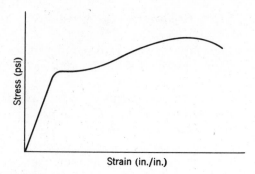

FIGURE 5.5.   Stress–strain curve for low-carbon steel.

an initial straight-line portion, and therefore the selection of a modulus value is relatively simple (see Figure 5.5). However, the rubbery materials have a continuously curving stress–strain plot, and hence the modulus slope in this case must be arbitrarily defined. At the present time, many laboratories will plot the stress–strain curve and draw a secant to the curve at some value of strain. The slope of this secant line defines the modulus. However, different laboratories use different strain values, so that the modulus value is hard for the consumer to interpret. Modulus values at 30, 50, 100, and even 300% strain have been reported in the literature. Figure 5.5a shows a stress–strain curve with a 50% modulus value reported.

The modulus of elasticity is affected by both specimen shape and testing rate. A flat specimen cut out of a pressed sheet is often used to determine the modulus. This is probably the best available shape for determining the properties of the rubber to be used in a preformed gasket, but is quite unrealistic in evaluating a bulk sealant. The sealant should be formed into

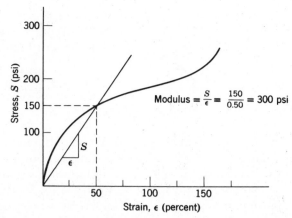

FIGURE 5.5a.   Typical stress–strain curve for an elastomeric sealant.

a ½ × ½-inch specimen, which is more representative of the actual shape of the sealant bead in the joint. Also, the faster the testing rate, the higher the modulus of elasticity. Most tests on the tensile adhesion specimen are run at a rate of separation of 0.2 inches per minute. The stress (psi) can be reported for 10, 20, 25, 50, and 100% strains (elongation) and compared after various exposures. Karpati uses this test specimen in many of her studies.

5.13. Abrasion Resistance

Test data on abrasion resistance are not reported for sealants. Yet this property is important where the sealant is to be used in expansion joints on terraces, patios, ramps, and garage decks. However, this is one area where architects have recognized the need for traffic-resistant sealants, and sealant manufactureres have made special sealants for this area of application. The sealants generally have a higher modulus. The major sealant specifications all include a class for ±12.5% movement capability, which was designed for traffic-bearing sealants, but in recent years most sealant manufacturers have derived sealants that will take ±25% for this area. Due to the fact that all major specifications contain both classes, the architect or specifier should make certain that the wording in his or her specifications spells out the proper class; otherwise a sealant with lower movement capabilty could be substituted. This test is run by exposing a sample to a rotating abrading disc and measuring the weight loss after specific times and treatment. Interpolation of the test data and proper technique are required for proper evaluation, since low-modulus compounds tend to gum up the disc and give erratic results.

5.14. Solvent and Chemical Resistance

Resistance to solvents or chemicals is not a standard requirement for sealants. Most areas of application would not require this property. However, the tests are easy to run and most sealant manufacturers will report resistance to various solvents and chemicals. Polysulfide sealants for integral fuel-tank sealing must have resistance to jet fuels, and several federal specifications are written for requirements in this area. In other applications around airports, some runways and work areas require expansion-joint sealants that are resistant to jet fuel spillage. Sealants may be used on plant and laboratory floors where some resistance to chemicals is required. The architect can select materials that will meet the particular need providing the environment can be determined in advance. The tests are easy to run, and usually require the testing of strips of sealant, measuring the length increase in the strip, and converting this figure to a volume change. The time for immersion varies considerably, and it should be recognized that the absorption of solvent into a sealant is not rapid. Consequently spillage and quick removal wil have no effect on a sealant. Immersion for 24 hours is generally required to pick up significant differences. For fuel-tank applications, the sealant is immersed in fuel for several days, and the fuel may be heated to 180°F—which can occur on parts of an actual storage area.

5.15.   Water Resistance

Adhesion of sealants to various substrates is affected by the presence of water. This is most apparent where the substrate may be saturated with moisture. Masonry saturated with moisture can cause salts to be slowly deposited on the surface or at the substrate. For this reason, primers are often used to prevent deterioration at the bond interface. The primer reinforces the surface and keeps water away from the sealant interface. Water immersion is a standard environment in most sealant testing. Long-time immersion, as in swimming pools, can cause ultimate loss in adhesion since the fillers can become hydrated, plasticizers can be extracted, or water can be absorbed by fillers, plasticers, and polymers. For this reason, most sealants are not recommended for continual immersion and the architect should design joints so that any trapped water can be weeped out. Water trapped in parapet walls can affect the adhesion of many sealants—and this is a common failing in parapets—where the top slab is absorbent and will allow moisture to gradually penetrate. Water accumulates within the parapet and then slowly oozes out to cause efflorescent streaks down the wall. Sealant that has lost adhesion generally leads to the gradual disintegration of the parapets. The top slabs must be nonabsorbent; if not, these surfaces must be treated with water repellents that will allow the walls to breathe but not absorb water. Metal caps have been designed to cover parapets and eliminate this problem area.

Latex-based sealants contain emulsifying agents that are used to keep the latex in a stable condition before formulation into a sealant. For this reason, latex-based sealants will absorb water and lose adhesion. This class of sealants is generally recommended for indoor use only.

5.16.   Toxicity

The toxicity of sealant materials is not a major problem. Since many of the modern sealants are sophisticated chemical compounds and since many contain solvents, these materials may be allergens. Continuous exposure may cause some skin or eye irritation. Consequently, as with most chemicals, it is advisable to use caution and avoid repeated or prolonged contact with the skin. Since allergy reactions vary so widely from person to person, there is no real test to define this property. Some sealants have contained toxic materials and because of the OSHA regulations, many sealants for use in the home have been reformulated to remove toxic chemicals where possible. If sealants still contain toxic ingredients, the containers must be visibly labeled to note poisonous compounds. Some one-part silicone sealants release acetic acid upon cure, which is generally not a problem unless one is working in a confined area. Other one-part silicone sealants may release complex amines. Here again this is not a problem unless confined. All sealants will contain precautions pertaining to use or ingestion of the sealant.

5.17. Electrical Properties

Sealants can be formulated for electrical resistance, but require different ingredients on occasion. Where the sealants are to be used in electrical environments, fillers, plasticizers, and curing agents are selected for optimum electrical resistance. Although sealants with fillers have been used in electrical applications, generally the clear epoxy or modified epoxy resins have been preferred, both for clarity and pourability.

5.18. Cyclic Tension and Compression Testing

The major sealant reference standards include cyclic tension and compression testing on test specimens. In addition, many manufacturers and other agencies have incorporated this test into their own evaluation process. Although most apparatuses are more than adequate, more important are the heat aging and pretreatment of the specimens. Some apparatuses have been modified to include compression cycling at higher temperatures and extension at lower temperatures. Other apparatuses also include simulated rain and ultraviolet exposure. The introduction of a number of test cycles in any 24 hour period is glamorous but not necessarily rigorous. If a sealant has not been exposed to any environment for at least 24 hours, it has not been affected to any extent. Specifically, for compression set to occur, the samples may require at least 24 hours compression at an elevated temperature for some materials to become modified. Also, there is a marked difference in the amount of water that may be absorbed in hours versus 1 or 2 weeks. Most sealants are not affected by several hours of exposure to water. When this has been extended to 1 or 2 weeks, the adhesion properties are the first to be affected.

Recommendations have been made to upgrade the exposures to water to 2 weeks' immersion and to toughen the requirements after exposure.

The following sections list a few of the testing devices that have been developed. Although this list does not attempt to delineate all the specialized pieces of equipment in use, it is broadly representative of the types of equipment that have been developed.

5.18.1. Bostic Tester

This testing equipment, built by Bostic, Ltd. of England, was one of the earliest apparatuses devised in the early 1960s for cyclic testing. It served as a model for later equipment that was developed for this purpose. The Bostic Tester is capable of testing both butt and lap joints because it incorporates both a tension and a shear set-up. The apparatus subjects the specimens to incremental movements controlled by time-limit switches. The capacity of the machine is six 2-inch long specimens at a time on each unit; and the machine is equipped with three units, two for tension-compression cycling and one for shear. The apparatus can be enclosed in an environmental chamber.

5.18.2. Applied Test Systems Tester

ATS has done an admirable job in designing equipment that will handle specimens for a number of specifications. Several models are available. The basic tester is comprised of four test stations with provision for additional test fixtures to permit testing of 8 or 12 specimens simultaneously. The testers are designed to provide cold and hot environmental testing. The testers have variable speed to accommodate the Japanese standards and can also test against U.S. Federal and ASTM standards (see Figure 5.6).

FIGURE 5.6.   ATS Tester. (Courtesy of Applied Test Systems, Inc.)

5.18.3. Aymar Tester

This tester was developed to accommodate the cyclic tests in federal specifications. The apparatus tests six test assemblies simultaneously and can be placed in an environmental chamber.

5.18.4. TNO Tester

This apparatus was developed by W. Van As at the Institute TNO for Building Materials in the Netherlands. The machine tests one specimen at a time in a tension–compression cycle.

5.18.5. Dominion Tester

This apparatus was developed by the Dominion Rubber Company of Canada. A compression–extension tester, it has a capacity of 24 samples, each 6 inches long. This apparatus, which includes a temperature chamber, applies an incremental movement to the specimens. The time switches give this machine great flexibility in programming the test cycle.

5.18.6. Sika Tester

Developed by the Sika Chemical Company, this machine tests six 2- or 3-inch-long specimens in a tension–compression cycle. Wheel mounted, it can be moved into a cold room for testing. The machine is built entirely of aluminum and stainless steel to prevent corrosion from condensation.

5.18.7. PCA Tester

This apparatus was developed by the laboratories of the Portland Cement Association. It is built to include load cells for measuring stress as well as strain. The apparatus tests 20 4-inch-long specimens.

5.18.8. ORD Tester

The testing machine developed by the Ohio River Division Laboratories of the Corps of Engineers is probably the largest sealant-testing apparatus in existence. This machine will test 120 2-inch-long specimens in a controlled environment.

5.18.9. Bureau of Reclamation Simulator

This apparatus was designed for testing canal sealants. The apparatus will test two 1-foot-long specimens that are partly immersed in water. This is an outdoor testing device, and the joint movement is actuated by black plastic rods that are affected by temperature changes.

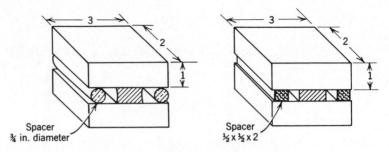

FIGURE 5.7.   Recovery test specimens.

5.18.10.   Thiokol Tester

The Thiokol Corporation apparatus cycles test assemblies simultaneously with temperature and cyclic movement. As the assembly is heated to 158°F, the sealant sample is compressed. Then in cooling down to −40°F the sample is extended. The samples are held at the extremes in temperature for a short period, and a complete cycle takes 24 hours.

5.18.11.   Creep, Stress Relaxation, and Recovery

The problems of creep and stress relaxation are intimately related. Both are time-dependent measures of the amount of flow in a sealant under load. Creep is the deformation with respect to time under constant load. Stress relaxation is the load with respect to the time under constant deformation.

Stress relaxation is sometimes described by the somewhat limiting term "compression set," and the recovery of sealants is generally reported at the end of a compression set test. Specimens are blocked in place at a fixed deformation and allowed to remain for a period of time. At the end of this

FIGURE 5.8.   Creep test of three sealant specimens.

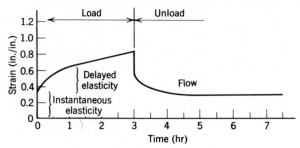

FIGURE 5.9. Creep curve.

period the specimens are freed and the percent of recovery is measured. Figure 5.7 shows a set of recovery specimens.

The sealant in the joint is actually in the stress relaxation situation; that is, the deformation is imposed on the sealant. The joint moves and the sealant must accommodate itself to this movement. When the sealant is held in the deformed position, the material begins to flow and readjust itself. All sealants will exhibit this phenomenom to some degree.

Stress relaxation or compression set is especially damaging to preformed gasket seals. This type of seal is compressed and placed in the joint in the compressed state. Consequently, if there is any appreciable amount of stress relaxation, the seal becomes ineffective.

Structural silicone sealants have very little stress relaxation, and consequently will show almost complete recovery after removal of an imposed load. At the other end of the spectrum, the solvent-based acrylics and butyl show very little recovery.

Paradoxically, although the sealant in the building joint is in the stress-relaxation situation, the creep property is much easier to measure. Figure 5.8 shows a typical creep test in progress. The creep of a sealant is generally reported as a curve of deformation versus time under constant load (Figure 5.9). The creep curve is valuable to the formulator because it indicates three

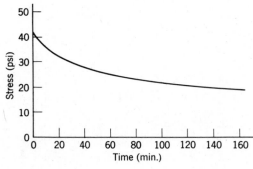

FIGURE 5.10. Stress-relaxation curve.

TABLE 5.1. Performance Capabilities of Generic Classes of Sealants

Generic Type	Parts	Maximum Movement Capability	Joint Limits W × D (in.)	Expected Life (yrs.)	Weight Shrinkage (%)	Resistance to		
						UV	Ozone	Heat Aging
Oil and resin base	1	±5%	¼ × ¼	10+	10	fair	fair	hardens
Butyl—mastics	1	±5%	⅜ × ⅜	10+	5 to 20	fair	fair	stays soft
Butyl—curable	1	±10%	½ × ⅜	10+	5 to 20	fair	fair	good
Butyl/polyisobutylene	1	±10%	tapes	20+	0	super	super	stays soft
Polyisobutylene	1	±10%	thin beads	20+	0	super	super	stays soft
Emulsion acrylic	1	±7.5%	½ × ½	10+	15	poor	poor	hardens
Solvent acrylic	1	±12.5%	¾ × ⅜	20+	10	good	good	toughens
Chlorosulfonated polyethylene	1	±12.5%	⅝ × ½	20	10	good	good	toughens
One-part polysulfide	1	±12.5%	¾ × ½	20	10	crazes	crazes	toughens
Two-part polysulfide	2	±25%	¾ × ½	20	10	crazes	crazes	toughens
One-part urethane	1	±25%	1¼ × ⅜	20	10	good	good	good
Multipart urethane	1	±25%	2 × ½	20+	4 to 10	super	super	super
Silicone—structural	1	±25%	¾ × ⅜	20+	4	super	super	super
Silicone—low modulus	1	over ±25%	¾ × ⅜	20	4	super	super	super

Precautions: Contact sealant manufacturer for priming instructions to various surfaces. The above capabilities are based on superior sealants meeting ASTM or federal specifications. Users should require certification from qualified laboratories. All joint depths to be controlled by back-up rod.

TABLE 5.2. Areas of Application for Generic Classes of Sealants

Generic Type	Areas of Application
Oil- and resin-base	Glazing of small wood and metal sash, small home repairs
Butyl-mastics	Non-moving masonry joints, accoustical sealants, home repair
Butyl-polyisobutylene	Curtain-wall interlocking joints, glazing, building tapes
Butyl-curable	Glazing over neoprene glazing gaskets, home repair, nonmoving masonry joints, metal ducts
Polyisobutylene	Preshimmed glazing and building tapes, primary seal for insulating glass
Emulsion acrylic	Indoor wall and ceiling joints
Solvent acrylic	Perimeter joints around doors and windows, medium-movement joints, toe and heel glazing beads, control joints, panel-to-panel joints, cap bead over tape, back bedding, bedding of mullions, panels, frames
Chlorosulfonated polyethylene	Perimeter joints around doors and windows, medium-movement wall joints
One-part Polysulfide	Building expansion joints for moderate climates, not for glazing
Two-part polysulfide	Same as one-part polysulfide
One-part urethane	Building expansion joints on building, not for glazing, expansion and control joints, precast concrete panel joints, tilt-up panel joints, curtain-wall joints, perimeter caulking of windows, doors, panels, beding of mullions, panels, frames, not for immersion
Multipart urethane	Same application areas as one-part urethane, and in addition, traffic joints, joints for moderate and tropical climates, some plaza joints, not for immersion
Silicone—structural	Glazing, structural glazing, butt glazing, metal and glass building joints
Silicone—low modulus	Metal and glass building joints, some masonry joints with special primers

distinct phases of sealant behavior. These are (1) instantaneous elasticity, (2) delayed elasticity, and (3) flow.

The stress relaxation of a material may be reported as a curve of stress versus time at constant deformation; or as a "stress relaxation time," which is the time required for the stress in the specimen to decay to 36.8% of its initial value (Figure 5.10). The stress-relaxation test is conducted by extending the sealant specimen 25 to 50% in a tensile testing machine. The specimen is extended as fast as the testing machine will operate. The machine is then locked into position with the sealant specimen at a constant deformation. Then the decay of stress with time (as the material relaxes) is plotted.

In the final analysis, what the consumer wants to know is how these properties affect the performance of the sealant in the joint. Unfortunately, neither the creep test nor the stress-relaxation test lends itself to data that the potential consumer can readily use. Consequently, the recovery of the sealant is the property generally reported. Recovery is the only measure of flow characteristics that is called for in any standard specification.

The inference in the recovery test is that only a high-recovery material makes a good sealant. However, there are many instances—such as heel beads for glazing and joints with irregular shapes—in which the low-recovery sealant is the better choice. The air seal in two-stage weatherproofing frequently indicates the low-recovery sealant as the better choice on a price per performance basis. However, any sealant selected must be able to take the movement that can take place. Each sealing job must therefore be handled individually.

Table 5.1 compares various generic classes of sealants in several pertinent properties helpful in design criteria. Table 5.2 gives application areas for the same generic classes.

6

Accessory Materials

Although the performance of a finished sealant may depend primarily on the properties of the base polymer, the performance is also dependent on the accessory materials. If the amounts of accessory materials are unreasonably high, then the finished sealant may have little or no resemblance to the original polymer. A sealant contains base polymer, fillers, plasticizers, thixotropic agents, adhesion promoters, catalysts, curing agents, and other ingredients. Various external accessory materials such as primers, bond-breaker tapes, back-up materials, and other supports are needed in order to effect a well-sealed joint.

6.1. Base Polymer

The base polymer for a sealant can be in any of three forms: a solid rubber, a liquid polymer, or a latex emulsion dispersion. A solid rubber requires solvent to reduce the mass to a liquid consistency. The major manufacturers can supply the rubber in a solution grade for small sealant manufacturers. The solid rubber is used to make solvent-release sealants, which are generally quite simple, consisting essentially of polymer, solvent, and fillers. This makes for simple formulation, few storage problems, and limits the areas of application. Polymers used in this category are butyl rubber, chlorosulfonated polyethylene, and various acrylic polymers.

Liquid polymers can be used to make sealants with little or no solvent or plasticizer. However, in most cases plasticizer and some solvent is used to modify both the handling properties and the cured properties. Liquid polymers used to make elastomeric sealants and caulks include polysulfide, various urethanes, silicones, polymercaptans, and polyisobutylene.

The latex emulsions are generally of three types: acrylic, polyvinyl chloride (PVC), or polyvinyl acetate (PVA). These emulsions are very easy to formulate since the starting base is a very fluid water emulsion—to which various fillers and other ingredients are added to give the finished sealants.

6.2. *Polymer Content*

There is an optimum polymer content for each polymer base. From a performance standpoint, there is an optimum balance of filler, plasticizer, and polymer to give the best performance possible with any system. Unfortunately, because of the high costs of polymer, various sealant manufacturers have to juggle the ratios of ingredients to make a profit. A manufacturer's sales policies, ethics, and liability all play a part in the final formulation. Since the base polymer is usually the most expensive component in terms of total cost, it is usually the first item to be trimmed. With each polymer covered, only typical formulations for good-quality sealants will be discussed. With most sealants, it is possible to water down the sealant with cheap fillers and plasticizers so that the polymer content could be considerably lowered. It also follows that low-polymer-content sealants will not meet high-quality performance specifications.

For solvent-release sealants, the acrylic rubber or chlorosulfonated polyethylene must be dissolved in a suitable solvent. One grade of acrylic rubber can be purchased in a 83%-solids solution that makes the manufacture of the sealant considerably easier. A good-quality formulation may have the following general composition:

Acrylic resin (100%)	35–40%
Plasticizer	2–10%
Fillers	45–55%
Solvent	10–15%

The plasticizer might be liquid at room temperature, or may require heat—in which case the sealant is a so-called "heat grade," which could lower solvent content and also lower the weight loss. Cure is mostly obtained by the evaporation of solvent. Too much plasticizer cannot be used, since it would cause the sealant to sag. The solvents are usually toluene, xylene, or blends of the two. The fillers are mostly ground calcium carbonate with small quantities of thickening agents and colorants. Because of the high solvent content these sealants are only recommended for medium-movement joints in the order of ±12.5% joint movement. Their solvent nature permit these sealants to wet most surfaces. Some acrylic polymers may have some cure on exposure to air and thus exhibit some recovery.

The latex emulsions are water-based supplied at 55 to 65% solids content. With suitable fillers and some plasticizer, they can be formulated into sealants that give good performance indoors where rain cannot affect the adhesion. A good-quality latex sealant might be formulated as follows:

Polymer (100%)	20–25%
Fillers	50–55%
Plasticizers	10–15%
Water	10–15%

The fillers are generally ground calcium carbonate along with small quantities of thickening agents and pigments. The plasticizer is compatible with the polymer and may be added as a fluid, or the polymer can be internally plasticized. The sealant can have considerable shrinkage, since a weight loss of 15% water results in a volume shrinkage of approximately 30%. However, when used properly these sealants can do an excellent job on plaster walls and other indoor surfaces. They are less expensive than the elastomeric sealants and are satisfactory for use in joint movements up to ±10%.

The polyisobutylene caulks can be based on the liquid polymer or the solid polymer in blends with the liquid. The solid polymer is used in making dual-seal insulating glass units where the sealant is extruded hot from an extruder that lays a firm round small bead on the spacer prior to laying the glass. The liquid polymer can be mixed with fillers to give noncuring caulks for use in curtain-wall interlocking joints or on metal building overlapping joints. Butyl rubber may be blended with the polyisobutylene to give sealants that may or may not cure upon exposure to air. Also these blends are used to make various tapes in either a cured or uncured composition. Some blends may enploy solvent and thus be a solvent-release system.

The first polysulfide sealants introduced to the building trade in the early 1950s were aircraft sealants for use in integral fuel tanks. These sealants had a polymer content of at least 80%, used carbon black as a reinforcing filler, and cured to a tough elastomeric sealant that gave adhesion to aluminum and was resistant to the action of jet fuel—while it kept all the aircraft joints sealed down to −65°F. Although these sealants performed satisfactorily in wing tanks, when tried as building sealants they failed miserably. They were too stiff, could not take much movement, and usually failed in adhesion because they were tough. It was apparent that to be satisfactory as building sealants, the polymer had to be modified with less reinforcing fillers, plasticizers had to be added, and the handling characteristics altered to make a good building sealant. Polymer content did drop and in some cases was too low. When the first sealant standard, ANSI A116.1 1960, was adopted, this was the first step to attempt to control polymer content by performance. A general formulation for a good-quality polysulfide sealant is as follows:

Polysulfide polymer	30–40%
Various fillers	30–40%
Plasticizers	20–25%
Curing agents	2–5%
Adhesion additives	1–3%
Miscellaneous	1–3%
Solvent	3–5%

The general fillers should be a blend of ground calcium carbonate and precipitated calcium carbonate, calcined clay to control pH, and some titanium dioxide for light colors. The plasticizers are mainly used to soften the cured polymer, and should be nonvolatile, non-extractible, and neutral. The curing agents can either be incorporated into the compound to make a one-part sealant or separated into a separate component for a two-part sealant. Adhesion additives are needed for every polymer system. Miscellaneous ingredients may include hydroscopic absorbers for one-part sealants, catalysts, retarding agents for cure control, and some solvent for ease in application, but this is sometimes abused.

Polyurethane sealants follow the same general guideline laid down for polysulfide sealants. Because of their greater popularity there has been a demand for many colors. About six standard colors are available in one-part sealants, and about ten for two- or multipart sealants. In several cases, the urethane sealants may be formulated a three-part sealants, in which case the base is a neutral shade and the third component is the color pack. This permits a very wide variety of standard colors, and offers the possibility of color matching to any desirable shade without the problem of package stability.

The unique thing about silicone sealants is that practically all organic type ester plasticizers are insoluble and incompatible with the silicone polymers. Consequently, the only type plasticizers are the silicone oils and fluids. Although they can be of various molecular weights, there is little cost advantage to using the lower molecular weights, which would be more volatile. Therefore most silicone sealants have very low weight loss after heat aging. The fillers are mostly silicas and not used in very high quantities. A typical formulation for a structural-grade silicone sealant is as follows:

Silicone polymer	65–75%
Silicone oil plasticizer	5–15%
Silica filler	15–25%
Curing agent	3–5%
Adhesion additive	1–3%

The lower-modulus silicone sealants would have a higher amount of silicone oil plasticizer and less cross-linking trifunctional silane complex.

The cross-linking agent gives the silicone sealants their high recovery when needed.

6.3. Fillers

The most common filler is calcium carbonate, because of its ready availability as a natural mineral, which makes it cheap. It comes in various sizes. The good grades of ground calcium carbonate have a particle size of 1 to 10 microns. A micron or micrometer is a unit of length one millionth of a meter. The cheap carbonates have a larger size, and larger amounts can be used in a sealant to reach the same viscosity. The precipitated carbonates have a particle size in the range of 0.1 to 0.3 microns.

The fillers for sealants need not be exotic. Since not much reinforcement is required, there is no major need for exclusively fine particle reinforcing fillers. On the other hand, the excessive use of coarse ground calcium carbonate is not desirable since it does not give any reinforcement and makes the sealant more sensitive to water. A good grind with some fine particle precipitated carbonate is a good combination for polysulfide and urethane sealants. Silica is needed for the silicone sealants. Titanium dioxide is needed whenever white or light shades are needed in a sealant. The use of some calcined clay may be desired for a better control pH. This is particularly useful if the polymer is sensitive to alkaline or high-pH fillers. Carbon blacks are only used in trace amounts as colorants. Aluminum flake was quite popular as a colorant in the past, but aluminum flake in powder form is active, which created problems on wet surfaces. In some cases the alkaline water runoff on masonry surfaces has caused reversion of some polymers sensitive to acidity, since the reaction of alkaline water with aluminum releases hydrogen gas that is acidic on the surface of the aluminum where the gas is formed.

In general, any high-quality pigment suitable for use in paint formulations may be used in compounding sealants. Studies have to be run to determine whether there is any chemical activity that can affect the polymer. Specifically, the iron oxides are not suitable fillers for most sealants because they are reactive. Fillers can comprise a greater volume in the sealants than the polymer. Fillers can be up to 90% of the volume in putties and oil- and resin-base caulks. Fillers in silicone sealants may comprise only 10 to 20% of the total volume. The other types of sealants fall in between.

6.4. Solvents

Solvents are needed in small amounts in many sealants to improve the gunability. Solvents are needed in greater proportion for the solvent-release sealants. However, in most cases it is possible to abuse the use of solvent and add more filler. The net result is a higher hardness sealant and more shrinkage when the sealant cures. The common solvents are toluene, xylene, petroleum spirits, and water. The consumer can spot the solvent content of a sealant quickly by reading the manufacturer's technical data sheets. A Material indicated at "86% solids" will have a weight loss of 14%, since solids content is usually expressed in weight. However, a 14% weight loss could become at 28% volume shrinkage, since the volatiles have a specific gravity of approximately 0.85 while the sealants will have a specific gravity of approximately 1.8 to 2.0, which means that for every 1% of weight loss there is a 2% volume shrinkage. Numbers can be deceiving and they hide the blunt truth.

6.5. Primers

Practically no polymer has adhesion to any substrate without the use of either adhesion additive, primer, or both. Many sealants are formulated so that they will adhere to some of the standard surfaces such as masonry, aluminum, and glass. Then if the sealant is used on another substrate, a primer may be required. Each supplier's literature will state the areas where primers are needed. The claims of various sealant manufacturers' literature vary considerably, and architects should be wary of claims for sealants that do not need a primer. It may follow that laboratory tests confirm good adhesion to virgin surfaces, but conditions on the job site are more variable.

All precast concrete is made in forms that have been swabbed with some kind of form oil to release the cured concrete from the form. These form oils vary from hydrocarbon oils and grease to metallic soaps. No one can predict how much contaminate has been absorbed on the surface of the precast member. Tests cannot be made in the laboratory, since the amount of material absorbed is very unpredictable, and the manufacturers that sell the mold release are reluctant to release either samples or composition. In many cases, primer to concrete is recommended as a means of preventing problems. Garden (11) states that; "Priming may be desirable in most instances if for no other reason than the fact that the joint surfaces are examined before installation of sealant."

Most metal surfaces can be contaminated with either a protective film or an oil left from the forming process. Metal surfaces can be purposely oxidized, such as the anodized aluminums, but this film is highly variable and definitely not a standard. Stainless-steel surfaces offer a very difficult challenge to most sealants. Softer metals and alloys such as lead, copper, zinc, and tern also create problems of adhesion. Today there are baked finishes on aluminum in the form of complex fluorinated polymers. Laboratory tests on factory production may be satisfactory in some cases, but everyone knows about lot-to-lot variation, and tests on the job site should be a normal follow-up.

In some cases, priming of masonry is required to first eliminate any contact of sealant with moisture. Moisture in concrete can become alkaline, which can affect sealant adhesion. Test in European standards require adhesion tests on concrete to be run in alkaline water. Thus the primer reinforces the surface of the concrete and seals off the surface while providing an easier surface for adhesion. Sealants in contact with two different surfaces may require a different primer to each surface. Also, most manufacturers supply their own primer. Primers from one manufacturer are not interchangable with primers from another.

Primers may be brushed, sprayed, or wiped onto the substrate. For masonry, a heavier primer film is required to penetrate the interface. For glass and metal, a very thin wipe that lays down a monomolecular film is all that is required. A heavy film could be completely unsatisfactory.

In some cases, a solvent wipe is first recommended to remove any grease or dirt. The solvent may contain an adhesion promoter, and this may be all that is required.

Sealant manufacturers offer different primers for various surfaces. There is no univeral sealant that will adhere to all surfaces without a primer!

6.6. Plasticizers

Plasticizers are necessary to reduce the modulus of a sealant. The general approach is to use a high-quality polymer that when cured gives good modulus and recovery, and then add plasticizer to maintain the good recovery while lowering the modulus. The best results are obtained by using nonvolatile plasticizers. However, the more volatile plasticizers are cheaper and consequently can be used to water down the polymer by incorporating more filler. The best control on volatile content is to tighten the requirements for weight loss after heat aging. A maximum value of 6% weight loss after 6 weeks' heat aging at 158°F would weed out those sealants using borderline plasticizers.

At one time the chlorinated biphenyls were widely used in most polysulfides as well as in other sealants. The biphenyls exhibited low volatility, had excellent compatibility, and were inexpensive. However, they were found to be cumulative in animal fats, and were eventually banned by OSHA. Even though banned, they were imported into the U.S. from Japan and other countries for several years. The doors were finally closed in the late 1960s. Plasticizers now used include phthalates, adipates, sebacates, phosphates, and other ester types. The field is quite open to most sealants except the silicones. Silicone sealants can only be plasticized by the silicone fluids and some selected solvents.

6.7. Curing Systems

The chemistry of cure is quite complex. Although there are innumerable ways to cure a polymer, the selection of the proper curing agent is based on attempting to get the best properties at a reasonable cost, and to achieve stability in the system. For polysulfides, lead dioxide is most widely used for building sealants in a two-part sealant. Calcium peroxide is selected for one-part polysulfide sealants because it becomes insensitive when the compound is dehydrated, yet becomes active upon exposure to moisture. The properties of one- and two-part polysulfide sealants are not identical because of the complex mechanism of cure. Manganese dioxide is used for polysulfide sealants used on insulating glass since these sealants are less affected by ultraviolet light.

One-part silicone sealants are moisture sensitive, and the by-products can be acetic acid, various complex amines, or phenol depending on the variations in the polymer terminal. Two-part silicone sealants can be cured by catalysts that contain hydroxyl or amine groups to give a complete cure where fast production is desired in manufacturing insulating glass units.

Urethane chemistry is extremely versatile, and many combinations are possible for one- and two-part sealants. The polymer can be manufactured so that it is quite sensitive to moisture and yet not release CO_2, which can be a by-product.

The only control on cure mechanism is by upgrading performance requirements—since in the end, performance is the main criterion for selection of a sealant.

As the one-part sealants cure, they tend to "skin over" and cure from the surface inward. Once a tight skin is formed over the surface of the sealant, the cure proceeds much more slowly. For laboratory testing the one-part sealants are force-cured at higher temperatures and 100% relative humidity, whereas in the actual building joint a $\frac{1}{2} \times \frac{1}{2}$-inch bead of sealant may take several weeks to cure. The size of the sealant bead is therefore quite critical in the selection of a one-part sealant. The single-part sealants may be good choices for joints up to $\frac{1}{2}$ inch wide. They are not suitable for wider joints, such as the $\frac{3}{4}$ inch or wider joints often used with precast concrete panels. On the other hand, two-part sealants cure simultaneously throughout the mass and can be used successfully in wide joints as well as in the smaller ones.

The shelf life of all sealants is important. In two-part sealants, fillers may tend to settle out with time and cake after excessive storage. This makes for difficult if not impossible mixing. One-part sealants, because of their cure system, are more seriously affected by long storage. Single-part sealants furnished in the standard caulking gun cartridge may begin to cure, making them practically impossible to ex'rude. Although the useful shelf life varies for the various sealants, it is probably safe to say that any material more than 6 months old should not be used in important work.

6.8. Back-Up Materials

The primary purpose of the back-up material in the joint is to control the depth of sealant in the joint, and thus insure the proper shape factor. Another purpose of the backing is to provide support or reinforcement for the sealant material in horizontal joints, such as in floors and patios. Depending on the type of construction, the back-up material may be already in the joint—for example, the plastic or cork board joint sometimes used in pavement construction.

The back-up material must be unaffected by any solvent contained in

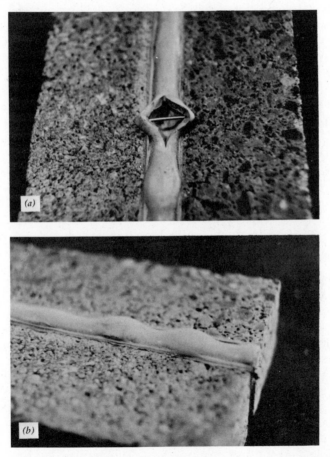

FIGURE 6.1. (a) Outgassing, which was caused by a puncture of a closed-cell backer rod. (b) Surface shot of (a) showing bubbling from a ruptured closed-cell backer rod. Workers must not use sharp tools to push rod into place. (Courtesy Backer Rod Manufacturing & Supply Co.)

the sealant. Back-up material containing asphalt, coal tar, or polyisobutylene should never be used. These extrudable oils are incompatible with some sealants and may cause loss of adhesion. It is also quite possible that these extrudables may cause staining of porous substrates.

Neoprene, urethane, polyethylene foams, cork, fiberboards, cotton rope, and jute have all been used as backing materials. The foams have been the most successful materials because they are quite compressible with very little spread. The foams are readily available in strip form in both round and rectangular cross-sections to fit most joints.

Rubber tubing made using neoprene, EPDM, or butyl, is an excellent but expensive back-up material. In essence, the tubing with the sealant acts as a dual system, since the tubing being partly compressed will also act as a water barrier in case the sealant may have lost adhesion.

Both open- and closed-cell back-up is used. The closed-cell back-up does not permit any water to contact the sealant, but has been a problem where the backer rod is first compressed into the joint, and then slowly releases gas into the uncured sealant if the rod is punctured during installation (Figure 6.1). This bubbling can become quite conspicuous if the wall is in the sunlight on a hot day before the sealant cures. The solution is to use open-cell foam. Open-cell foam is desirable if a one-part sealant that requires moisture activation is used. Here, the sealant will begin curing from both sides. However, with time the open-cell foam may become saturated with moisture and then cause adhesion problems or even frost formation and spalling. Properly vented walls would reduce this problem. There is no ideal solution, and the selection of back-up will depend on sealant used and job-site conditions.

6.9. Adhesion Promotors

Almost all sealants require an adhesion additive that results in the adhesion of a sealant to a substrate. Quite often, this additive may be a complex silane that is very expensive, but only a very small percentage is needed. The proper additive in the right amount can make or break a sealant. The selection of adhesion additives is by trial and error, and testing with various surfaces and in various environments. Additives may be selected for use on specific surfaces and may be different for one- and two-component sealants. Certain phenolic resins have been found very satisfactory for polysulfide sealants used on buildings. The silanes are now quite widely used in many sealants, particularly when the sealants are used on glass such as the insulating glass sealants. Obviously a wide variety of adhesion promoters is possible, and the only criterion for selecting one is performance. Sealant manufacturers must consider other problems such as package stability and proper incorporation into the sealant. The use of silane requires that the compound be essentially moisture-free, since silanes react in the presence of moisture on fillers and may even react with hydroxyl groups on various surfaces to get a chemical bond. The nature of adhesion is quite complex, and may involve direct reaction to the surface or establishing a polar bond. In any event, the proof is in the pudding, and all sealants need adhesion promoters.

6.10. Release Agents

In some joint design, a back-up plate is needed to support the sealant in a wide joint such as a garage ramp. In such cases, the joint could be 2 to 10 inches wide, and support in the form of a steel plate that is bolted to one side is needed. In such instances, three-sided adhesion of the sealant must be prevented by the use of a bond-breaker tape placed on the support plate. These tapes are standard items with tape manufacturers, and even masking tape will do an adequate job in an emergency. The polyethylene foam back-up rods are not a problem since sealant does not adhere to polyethylene. Where sealant is placed as a cap bead over an uncured butyl tape, there is also no problem of adhesion. Sometimes a joint is widened at the surface by grinding out some masonry. The original joint may have only been ¼ inch and may be failed. The solution is to widen the joint and the depth to ½ inch. Since adhesion is not desired at the bottom of the joint, then a bond-breaker tape is laid at the bottom of the joint to prevent three-sided adhesion. The sides of the joint will probably need priming.

6.11. Lubricant-Adhesives

The lubricant-adhesives are used only with preformed gasket seals. Depending on the type of installation, the lubricant-adhesive may serve only as a lubricant or it may serve both functions. In a typical glazing operation a simple soap solution may be used to ease the precompressed seal into the joint. However, in highway pavement joints—where preformed seals are widely used—both lubricant and adhesion are desirable. The lubrication function is necessary in order to place the seal in the joint properly, and the adhesion function is desirable in order to help the seal maintain its proper position in the joint as the pavement moves.

Considering the two functions, it becomes apparent that different materials are used as lubricant-adhesives. In applications that require only a lubricant, a soap solution or other nonoily film may be used. Where some degree of adhesion is desired, a thin film of any compatible sealant such as a neoprene may be used.

6.12.   Thixotropic Agents

Sealants applied in vertical joints must not sag even though they must have easy gunability. In most cases a small amount of solvent that flashes off very fast is a first step to improving non-sag properties. The use of special additives such as fumed silica imparts a thixotropic action or thickening, without appreciable viscosity increase if used in small amounts. Some precipitated calcium carbonates are coated with stearic acid or other surface-active agent that imparts thixotropic action. In the past filament fibers of asbestos were used in some one-component sealants. Some glass fibers may be used in epoxy compounds since they do not detract from transparency. The fibers may be used in other applications for vertical application of fluid-applied membrane.

6.13.   Miscellaneous

Chemical reactions are the basis of elastomeric sealants that convert to a rubber. As with most chemical reactions, an acidic or alkaline environment may be desired to either accelerate or retard the reaction. Two-component sealants may have a catalyst in one component that provides the desirable activity or control. There may be UV absorbers incorporated into the sealant as well as ozone inhibitors that are added if the sealant has undesirable sensitivity to these environments. Sealants are complex compositions, and knowing their overall capability and limitations will go a long way to proper selection of sealant in any environment.

7

Installation

Various types of sealants require different installation techniques. Sealants can be supplied to the job site in several forms, each of which require special technique and consideration. The types are as follows:

Single component.
Two component.
Tapes.
Preformed gaskets.
Foams.

7.1. Sealants

The one- and two-component sealants differ in both storage and preparation of the sealant for use, but the basic installation procedure is the same for both types.

First the joint is cleaned. This may require wire brushing, a dry wipe, or a solvent wipe. The back-up material is then placed in the joint. Many back-up materials do not create the problem of three-sided adhesion, since sealant does not adhere to materials such as polyethylene foam rod, and there is no problem if the sealant is laid against lightweight open-cell urethane foam rod. However, if the back-up is a solid such as fiber board, masonry, or metal, then bond-breaker tape is laid down to prevent three-sided adhesion of the sealant. If a neat job is desired, where the substrate is white marble and the sealant is darker, then masking tape may be placed on the substrate along both exterior edges of the joint. Many masonry primers are amber in color because they are based on phenolic resins, and

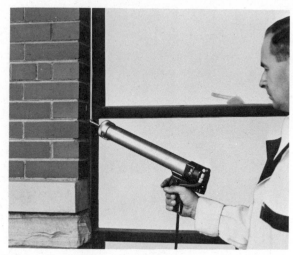

FIGURE 7.1.   Sealant application using an air-operated caulking gun. (Courtesy of Pyles Industries.)

these primers might stain the substrate if not carefully applied. This may be another reason for using masking tape. The joint is then primed, and allowed to dry, which requires only 5 to 15 minutes. The sealant is then applied from hand or air-operated caulking gun. Figure 7.1 shows a sealant being installed with an air gun. The surface of the joint is then tooled. Tooling forces the air from the sealant, forces the sealant into the joint to insure wetting of the substrate, and provides the proper contour to the sealant surface. The tool or gadget used for tooling may be a shaped tongue depressor, which may be dipped in a solvent or suitable soap solution to make tooling easier. The soap solution should be one that does not later cause a sealant stain. After tooling is completed, the masking tape along the sides of the joint may be removed.

There is no proper way to use a caulking gun. There are two schools of thought—the pushers and the pullers. Either method is satisfactory as long as the sealant is properly applied. More important is the cut angle on the nozzle and the angle of the gun to the joint. When properly applied, the sealant is forced against the three sides of the joint, and the surface is smooth and curved. Tooling is desirable to force out small air pockets. Tooling becomes necessary when the edges of the joint are irregular, as with some porous granite or poor precast concrete, and it becomes almost impossible to lay a smooth bead.

Weather is also an important factor in joint seal installations. All one- and two-component sealants cure very slowly at temperatures below 45°F. One-component sealants may take several months to reach a partial cure in cold winters. This may not be a serious problem since in the winter the joints are not moving much and may be frozen in place. Two-component sealants will cure faster than one-component sealants in the winter, but here again several weeks may be required for a fair degree of cure. The higher the ambient temperature, the faster the cure. Also the absolute humidity becomes higher at higher temperatures. Warm, humid climates accelerate one-component cures. If one-stage weatherproofing is being employed in a concrete building, rain can slow down a sealing operation for several days, since a satisfactory method has not been found for priming or sealing saturated concrete that may have 15 to 18% moisture content. In many cases, standing water accumulates in horizontal joints, thereby making sealing impossible.

7.1.1. Single-Component Sealants

Single-component sealants are furnished to the consumer in 1/10-gallon cartridges, or in 1, 2, or 5-gallon pails. The cartridges are clean and easy to handle, store, and use. Cartridges are simply inserted into a manual or air-operated caulking gun, the plastic tip is then snipped off, the barrier seal is broken with a long nail, and then the unit is ready for use.

Bulk material in the larger pails is applied with a bulk-loading caulking gun. The bulk-loading gun may be either a suction type, which is loaded by dipping the nozzle into the sealant and withdrawing a plunger, or a rear-load type, which may consist of a follower plate forced against the sealant causing the sealant to flow up into the cartridge through the adapter.

Figure 7.2 shows an air-operated caulking gun adaptable for either cartridge or bulk-loading. Special equipment for power-operated bulk loading

FIGURE 7.2. Air-operated gun adaptable for either cartridge or bulk loading. (Courtesy of Pyles Industries.)

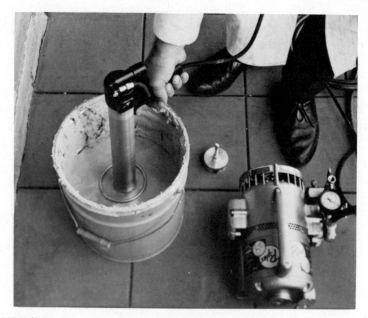

FIGURE 7.3.   Power-operated bulk loading. (Courtesy of Pyles Industries.)

is shown in Figure 7.3. Figure 7.4 is a portable air unit the caulking gun operator can conveniently carry onto a scaffold or to isolated job locations. The air unit is strapped to the operator's body, leaving both hands free for work.

The advantages of single-component sealants are that they require no on-site mixing and maintain a longer working life. This longer working life is of value, especially when the contractor is using material furnished to the job in the larger containers. The contractor may open a 5-gallon pail of sealant and refill guns from it over a period of several hours. It is also possible to open a bulk container one day and still use the material the following day. There may be some skinning over of the material in the opened pail, but this skin may be removed and discarded, and work can proceed. However, since one-component sealants require several days to cure at warm temperatures, they should not be used in pedestrian areas or on sidewalks.

7.1.2.   *Two-Component Sealants*

Two-component sealants are installed into the joint using the same types of caulking equipment as for the single-component sealants. An additional requirement of the two-component materials is that they be thoroughly

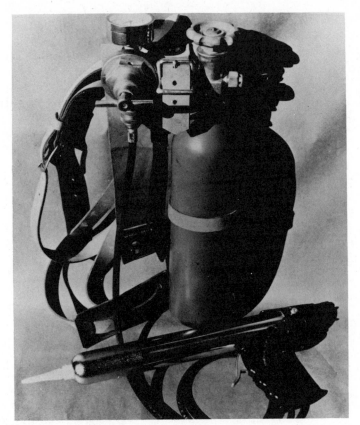

FIGURE 7.4. Air gun with portable air supply. (Courtesy of Semco Sales and Service.)

mixed at the job site. This is not too difficult, since the base and catalyst are usually of different colors, and unless the sealant is thoroughly mixed it appears streaked. The two-component materials may be furnished in the same container sizes used for the single components, but the larger percentage of material is supplied in 1, 1½, 2 or 5-gallon pails. These materials are always supplied to the job site in premeasured quantities, so that if all the material in container A is mixed with all the material in container B satisfactory proportioning of components is assured.

Some sealants are supplied in a three-component system, namely a base, catalyst, and color component. This permits a wide selection of colors and color matching of various substrates. The great bulk of or multicomponent sealants is mixed at the site, using a paddle on a slow-speed electric drill of approximately 450 rpm. Predesigned paddles are available from the sealant manufacturer. Faster mixing—which whips excess air into the mix and also heats up the sealant—would shorten working life and cause more problems in eliminating air after caulking.

Excellent construction-site mixers are available that use a rotating screw to mix the materials. These mixers, although expensive, do a good job of mixing without entrapping air in the sealant. After mixing, the apparatus (see Figure 7.5) ejects the material so that it can be conveniently loaded into bulk-type caulking guns. Several European companies have designed a screw that can be worked into a cartridge containing both components, and the sealant is mixed in the cartridge. This eliminates transfer of material into a cartridge. The process requires putting two components into the cartridge, and the catalyst is added last. Surprisingly, there is no curing at the interface of the two components. Although popular in Europe, the idea has never caught on in the U.S.

Good-quality two-component sealants are generally cheaper than one-component sealants, which comes as a surprise to many people. The general expectation is that one-component sealants should be half as expensive as two-component sealants, but cartridges are expensive and the labor to fill them is also expensive. A good quality cartridge must be impermeable to moisture. Specifically selected and coated plastic cartridges are generally required when the sealants are highly sensitive to moisture.

Two-component sealants can have a work life that varies from 1 to 8 hours, so that this factor must be considered in warm climates where the work life could get as low as 1 hour at 100°F. Because of fast through cure, two-component sealants are desired for pedestrian traffic areas. Two-component sealants are generally better in quality, since in all cases the cure mechanisms are different. Specifiers are too concerned about the possibility of a poor mix of a two-component sealant; instead, they should be more concerned about the quality and performance of the sealant they select.

Two-component sealants used in highway and airfield joints may be installed with special equipment. The two components of the sealant are furnished in a 1:1 volume ratio. They are pumped separately through hoses, thoroughly mixed at the nozzle, and extruded into the joint. Then this entire unit is truck or trailer mounted in order to cover the distance involved in pavement sealing.

In recent years, formulators have derived a one-component hot-melt sealant for use on airport runways, highways, and canals. These hot-melt compounds based on PVC are extruded into place when heated to a pumpable state at approximately 300°F. They cool immediately, and become elastomeric upon cooling. A number of ASTM standards have been derived covering these newer materials, which can be applied with much simpler equipment. Clean joints are also required here. Very recently, one-part silicone low-modulus sealants have been tested and proposed for highway joints, due to their high movement capability. Test trials have attested to this very new field of application for silicone sealants.

An alternate system for furnishing two-component materials in cartridges is to premix the components at the factory, load the sealant into polyethylene

FIGURE 7.5.   Mixer for two-component sealants. (Courtesy of Semco Sales and Service.)

cartridges, and quick freeze the sealant to stop the curing process. This process is used by the aviation industry to control the preparation, testing, and application for sealant used in integral wing tank sealing. The cartridges when filled are immediately immersed in an acetone-dry ice bath where they are frozen to −90°F in seconds. The cartridges are then thawed at the job site and used. The 6-month shelf life limitation is especially important

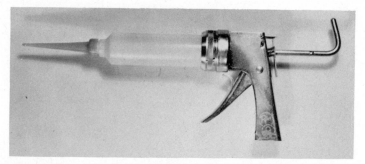

FIGURE 7.6. Handgun for use with premixed and frozen sealants. (Courtesy of Pyles Industries.)

with this type of packaging. Figures 7.6 and 7.7 show the hand and air guns often used with the frozen cartridges. One company that supplies this cartridge to the aircraft industry introduced this technique to Australia, where it was widely used during the 1960s. The practice has since declined for building use, but it is interesting to note the trends in various parts of the world.

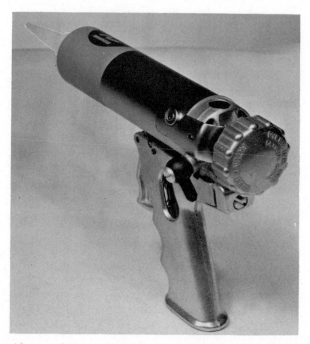

FIGURE 7.7. Air gun for use with premixed and frozen sealants. (Courtesy of Pyles Industries.)

7.2. Tapes

Tape seals to put it very simply, are mastic-type sealants that have been thickened enough to be furnished in roll form. They are generally supplied with a paper backing that is removed immediately prior to installation. Tapes come in round, square, and rectangular shapes. Tapes for metal buildings and other areas where the joint can vary quite widely are usually softer, more compressible and thicker, which enables the tapes to take a wide variation in width.

Where tapes are to be used in glazing, they are supplied in more accurate dimension. Tapes used for small lites may be non-shimmed where the united inches are less than 75 inches. The addition of one length and one width in the lite gives "united inches." For lites above 75 united inches, one of several types of pre-shimmed tapes are available. Since the shim in the tape is a cured butyl rod, the tape is made a little wider than the shim in the tape so that the surface can be properly wetted by the sealant. The standard unshimmed tapes come in thicknesses ranging from $\frac{1}{16}$ to $\frac{1}{4}$ inch and widths of either $\frac{3}{8}$ or $\frac{1}{4}$ inch. The pre-shimmed tapes come in thicknesses from $\frac{1}{8}$ to $\frac{1}{4}$ inch and widths of either $\frac{3}{8}$ or $\frac{1}{2}$ inch. The shim is usually $\frac{1}{16}$ inch less in diameter than the thickness of the tape, thus allowing $\frac{1}{16}$ inch for squeeze down and proper wetting. The tapes are generally solvent-free and acquire their softnes from the fluid polyisobutylene that may be used in conjunction with butyl rubber depending on the viscosity or plasticity desired.

Although joint cleaning is not critical, it is desirable to remove any oil or dirt that would prevent good wetting. Priming is seldom necessary due to the good wettability of the tapes. The actual installation of the tape sealant is accomplished by hand. Strips of tape of the proper length are cut from the roll with a knife or scissors, and the sealant is pressed into place. Butt splices and corner splices are easily formed by pushing the tape into place with the fingers. Common practice also calls for gunning some sealant over the splice areas, in order to prevent future leakage in these areas. Because of the types of application, joint tooling or finishing is seldom necessary.

7.3. Preformed Gaskets

The preformed gasket seals are used in literally hundreds of specialized applications and consequently are supplied to the job site in a multitude of shapes. Common to all these shapes is the fact that they are installed under compression. Since adhesion is no problem, joint cleaning may consist of simply blowing or brushing loose dirt out of the joint. A lubricant is wiped along the gasket, and then the gasket is pressed into the joint. A simple device may consist of a lever arm with the pivot attached to a suction cup on the glass. Such a device may be needed since the gasket is used at times to press against the glass, which is against another material such as a tape or a foam gasket.

The preformed gaskets may be supplied in a rectangular shape if straight stops are used. The preformed gasket can have an adhesive on one side to hold the gasket in place until the panel or lite is put into place. The gasket may also have adhesive on both sides. For small lites, gasket with adhesive may be sufficient. For larger lites it may be necessary to use a cap bead, since the gasket will work itself loose with wind buffeting and alternate expansion and contraction of the panel or glass lite. Where the stops are extruded with special designs, there may be openings into which shaped gaskets with keys may be locked in place (Figure 7.8). In exterior joints, such as with precast concrete or metal curtain-wall panels, the building panels are erected first and then the seal is compressed and placed in the joint. In other applications where the precast is placed against a steel member, the member may be slotted, a gasket keyed into the slot, and then the panel placed against the gasket.

In a glazing application the installation will vary according to the type of window unit, but the following sequence is typical whether glazing from the inside or outside:

1. Place a tape or foam gasket on the window framing.
2. Set the glass next to the gasket or tape.
3. Fasten the inside or outside stop.
4. Insert a dense gasket.
5. Cap bead on outside or inside if desired, if straight-sided gaskets are used. Cap bead may be used also on large lites even though keyed gaskets are used.

The critical factors to be considered in gasket installation are stretching and splicing. The gasket must be designed for the joint. Where foam gaskets are used, compression of up to 40% is necessary to form a seal. This compression is generally applied by the second gasket or spacer. Stretching to insert the gasket is undesirable, since the stretching decreases the cross-section and reduces the interface pressure to function properly. The gasket sections must be carefully cut to length for installation because there is no adhesive that will form a satisfactory structural splice.

The exterior gasket, whether foam or dense, may have molded corners so that the gasket is now a picture frame. Most gaskets can leak at corners unless properly spliced or molded. The gaskets must be exactly sized and are specifically designed to be ⅛ inch longer in each dimension than the actual size. This puts the gasket under a little compression when in place. Molded gaskets prevent leakage at the corners. Molded picture-frame foam gaskets may also be placed on the inside. If four pieces of dense gasket or tape are used on the exterior side, then a sealant is usually gunned on the corner splice and 2 to 4 inches up each leg—to insure a sealed corner—

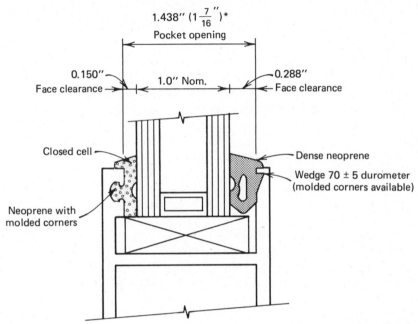

FIGURE 7.8. Typical compression glazing system. (Courtesy of Tremco, Inc.)

before placement of glass. In some instances, a heel bead or toe bead is gunned in place as a precaution to prevent wind-driven rain from infiltrating the building.

No joint finishing is required with these denser extrusions, because the gaskets furnish an excellent finished appearance. The use of a dry/dry gasket system applied from the inside makes for rapid, complete installation in one step, which is labor-saving. The essential requirements are weep holes when dry gasket glazing techniques are used.

7.4. Foam Gaskets

Closed-cell foam gaskets have at least as many possible applications as the denser gasket extrusions, and consequently are available in a wide variety of shapes. Foam seals are also a compression-type seal, and the installation process is much the same as for the denser extrusions. There is considerable overlap between foam sealants and preformed closed-cell gaskets. It might suffice to say that preformed gaskets are a higher-quality product and made to more stringent performance requirements, but some of the foam sealants will also apply. Foam sealants are supplied in a wide variety of densities, from 6.5 to 16 lbs per cubic foot, and a wide range of hardnesses from 15 to 40 on the Shore 00 scale, which is used to measure foam hardness.

Foam sealants can also be supplied with adhesive on one or both sides. Required cleaning is minimal, since the spongy nature of this type of seal permits it to flow into irregularities in the joint face. Priming of the joint is not required. The seal should be installed under a compression of approximately 25%. Compression beyond 50% is not recommended, since this high compression distorts the foam structure and causes a drop in the pressure the seal exerts against the joint wall. Foam seals should not be stretched longitudinally during installation, because this reduces cross-section and consequently reduces sealing pressure. Butt splices can be made satisfactorily in the field using a suitable adhesive. No tooling is required.

7.5. *Joint Cleaning*

Proper cleaning of joints is undoubtedly the most tedious step in the entire sealing process, but its importance cannot be overstated; no sealant will adhere to a dirty joint wall. Joint cleaning is somewhat less critical with preformed shapes, but it is still required. Loose paint, scale, or rust on metal surfaces, laitance on the face of concrete joints, and general construction dirt must be removed before the joint can be sealed. Sound metal surfaces such as new aluminum can be cleaned with a nonoily solvent and a soft rag. Scale, rust, and laitance are best removed by sandblasting. A clean wire brush can also be used for this purpose. In resealing work, old sealant can be cut out with a knife, and the remaining traces of old sealant can be removed from the joint faces by sandblasting or wire brushing. After sandblasting or wire brushing, the joint faces should be dusted clean with a soft brush or rag, or by vacuum. Compressed air should not be used to blow out the joints because oil from the air compressor is often sprayed onto the surface. The degree of cleanliness required by mastic sealants is almost in direct proportion to the amount of recovery in the sealant. Silicone and urethane sealants require very clean joints; whereas, at the other end of the spectrum, acrylic latex caulks and solvent-based acrylics require much less cleaning.

Aluminum mullions and sash, as furnished to the job site, present a special cleaning problem. Aluminum sections are often given a special lacquer finish at the factory. This lacquer coating should be removed so that the sealant can adhere directly to the aluminum. A strong solvent, such as toluene or xylene, on a clean rag can remove the lacquer.

7.6. *Heaters*

At one time most solvent-base acrylic sealants required heating to 120°F before they could be extruded satisfactorily from a caulking gun. Heating can be accomplished in a special heating chest available from the sealant manufacturers. Some contractors have built their own heating units, which are often nothing more complicated than a wooden box with a light bulb for a heating element.

Solvent-based acrylics have been modified, and today most sealants do not require heating at application temperatures of approximately 75°F. However, the easier extrudability has been obtained by using a different base polymer and probably a little more solvent. Some applicators still swear by the older product that required heating.

The same type of chest is sometimes used to thaw premixed and frozen cartridges of two-component sealants, but a bucket of warm water did a quicker job. Immersion heaters are also available to heat bulk sealant in 1 to 5-gallon bulk quantities when the sealant is too cold, and quick heating is required to enable mixing of catalyst and base. Heaters must be used with great care, since heat can accelerate the cure of the two-component sealants. If too much local heat is applied, some of the material may begin to cure before the rest of the sealant in the cartridge has had time to thaw out to a gunable consistency.

In the machine application of two-component highway pavement sealants, special heaters are sometimes used to control the viscosity of sealant components so that they can be easily pumped through the hose to the nozzles.

In some highway and airfield applications, hot poured rubber asphalt sealants are used. These materials require special double-wall melting kettles to heat the sealant to about 375°F. Other more recent compositions are the hot-melt vinyl compounds, which require the same type of treatment.

8

Polysulfide Sealants

8.1. Introduction

Polysulfide sealants were the first elastomeric joint sealants to be used in building construction in the early 1950s. They enjoyed a meteoric rise in popularity through the 1960s and early 1970s, but shared the market with solvent acrylics and urethanes in the mid-1970s, and with silicones in the late 1970s and early 1980s. Although there were a few earlier installations, the use of polysulfide sealant to recaulk Lever House (Figure 8.1) in the early 1950s signaled the acceptance of polysulfide sealants for the modern curtain wall. During the polysulfide boom, Thiokol introduced a "tested and approved" program that endorsed polysulfide sealants against Thiokol's own specification. A number of polysulfide reference standards, both domestic and international, helped the situation greatly. The first sealant standard in the U.S. was A116.1 1960, which became the forerunner of many standards throughout the world. Thiokol discontinued its "T and A" program in the late 1970s, probably due to reduced sales and the increased cost of running the program (which required considerable testing). The use of polysulfide sealants for building use has decreased for several reasons. First, the polymer was solely supplied by Thiokol, and was always expensive. Second, urethanes and silicone sealants exhibited better resistance to UV and ozone, and the urethane polymer could be easily made by sealant manufacturers or was readily available from suppliers. Third, the competetive pricing resulted in poorer quality by some manufacturers. Polysulfide sealants can still be made in good quality, but at a higher price. Polysulfide sealants are still the dominant sealant for making insulating glass units and will probably hold this position for some years. Hot-melt butyls have edged

FIGURE 8.1. Lever House, New York City. One of the first major buildings recaulked with polysulfide sealant in 1954. (Courtesy of Thiokol Corporation.)

into the picture, and silicone/butyl have now been recognized as the top-quality sealant combination for insulating glass but at a premium price.

Polysulfide rubber Thiokol A was first introduced by Thiokol Chemical Corporation in 1929. This rubber was supplemented by a number of other polymers, which led to the introduction of Thiokol ST in 1942, a mercaptan terminated polysulfide rubber. Liquid polymers that were a lower molecular weight version of Thiokol ST were introduced to plasticize ST but soon became prominent on their own as curable liquid polymers for use in the aviation industry. Panek (12) covers the complete development of the Thiokol polysulfide rubbers, latices, and liquid polymers along with cure formulations and chemical properties.

The number of formulators for building sealants during the boom was as high as 35, but has since been considerably reduced. Although the patents covering polysulfide liquid polymer manufacture ran out in the late 1970s, there have been no other manufacturers in the U.S. There are manufacturers of liquid polymer in Japan and East Germany that cater to some foreign markets.

Several manufacturers have introduced mercaptan terminated polymers with different backbones that have been labeled as polymercaptans as well as polysulfides. These polymers will be covered in Chapter 15.

8.2. Chemistry

Polysulfide liquid polymers are easily made, and a series of polymers was introduced over the years. The typical reaction is designated as follows:

$$nClC_2H_4OCH_2OC_2H_4Cl + nNa_2S_{2.25} \rightarrow$$
$$HS[C_2H_4OCH_2OC_2H_4SS_{2.25}]n\!-\!C_2H_4OCH_2OC_2H_4SH + nN_aCl$$

Dichloroethylformal is slowly added to a solution of sodim polysulfide plus emulsifying agents and a latex is first formed. After the latex is washed clean, then splitting salts consisting of sodium sulfhydrate and sodium sulfite are added, and the mixture is then acidified and coagulated and washed clean. The amount of sodium sulfhydrate determines the molecular weight of the liquid polymer. The number of repeating units determines the final molecular weight of the polymer. The use of small amounts of the cross-linking agent trichloropropane added to the formal introduces trifunctionality, which is desired to improve compression set resistance. The cross-linking agent also increases modulus and decreases elongation. Table 8.1 lists some of the various liquid polymers that have been made over the years for use in building sealants and other areas.

LP-31 has been used primarily for integral fuel tank sealing. LP-3 and LP-33 are used mainly as binders for solid propellant fuels, along with other polymers not listed. LP-32 and LP-2, and to a certain extent LP-12, are the main polymers for building sealants.

TABLE 8.1. Polysulfide Liquid Polymers

Polymers	Mol. Wt.	Cross-Link %	Viscosity, Poise	Physical State
LP-31	8000	0.5	800–1400	viscous fluid
LP-32	4000	0.5	350–450	slightly viscous
LP-2	4000	2.0	350–450	slightly viscous
LP-12	4000	0.1	350–450	slightly viscous
LP-3	1000	2.0	7–12	fluid liquid
LP-33	1000	0.5	7–12	fluid liquid

The cure mechanism is a simple oxidation of the mercaptan groups to form the cured polymer:

$$HSRSH \ [O] \rightarrow RSSR + H_2O$$

With a metallic peroxide MO_2 the peroxide is reduced to a simple oxide and the cured polymer formed. PbO_2 is most commonly used for fast curing builing sealants, and gives a good polymer. A second undesirable reaction that takes place in part is the formation of a metallic mercaptide.

$$HSRSH + MO \rightarrow RSMSR + H_2O$$

The amount of metallic mercaptide formation is a function of the activity of the original peroxide. Lead peroxide, PbO_2, gives good properties for building sealants. Manganese dioxide gives better heat resistance and resistance to UV and is used for aircraft sealants and for insulating glass. Calcium peroxide is necessary for one-component sealants, since it is inactive when the compound is dehydrated but becomes active upon exposure to moisture in the air. The inorganic chromates are also used as oxidizing agents since they also give better heat resistance for use as aircraft sealants. Organic peroxides can be used, but give very different properties, and such compounds are very difficult to get to adhere to various surfaces. The complete chemistry of curing requires additional trace catalysts and an understanding of the complex interaction of the various adhesion additives, fillers, curing agents, and by-products from the cure mechanism. The liquid polymers can also react directly with epoxy resins to give more flexible products with unique properties.

8.3. Compounding

LP-2 and LP-32 have the consistency of honey. These polymers cured by themselves using a paste of lead dioxide give very poor properties and would also be prohibitive in price to use as straight polymers. Therefore the use of fillers and plasticizers is desired to improve physical properties. First, the fillers improve the physical properties, but then plasticizers are needed to lower modulus and hardness. Both fillers and plasticizers also greatly reduce the cost of the sealant. A good basic formulation shown in Table 8.2 would have approximately 40% polymer content.

TABLE 8.2. Basic Sealant Formulation

LP-32	40% by weight
Plasticizers	20% by weight
Fillers	35% by weight
Miscellaneous	5% by weight

TABLE 8.3.   **Effect of Filler and Plasticizer on Physical Properties**

LP-32	100	100	100	100
Filler	—	80	—	100
Plasticizer	—	—	50	50
PbO_2 paste	15	15	15	15
Physical properties				
Tensile psi	100	500	50	300
Elongation %	350	400	1500	650
Hardness, Shore A	30	50	10	30

The miscellaneous items in Table 8.2 would include curing agent, activators, retarders, adhesion additives, and even dehydrating agents when necessary.

Satisfactory fillers would include a good grind of natural calcium carbonate, some good grind of calcined clay, some precipitated calcium carbonate, some titanium dioxide for whiteness, and thickeners. Plasticizers could include dioctyl phthalate, other high molecular weight phthalates, phosphates, adipates, azelates, and other organic esters—all having low volatility. The effect of filler and plasticizer is shown in Table 8.3.

The use of filler alone gives high tensile strength, but very high hardness. The use of plasticizer alone gives very high elongation, very low hardness, and very poor tensile strength. A proper balance of polymer, filler, and plasticizer gives a good balance of physical properties and satisfactory hardness. The use of nonvolatile plasticizer would result in a good retention of physical properties after heat aging as well as low volatile loss in sealant, which is needed for sealants that are intended to perform satisfactorily for at least 20 years.

Fillers and plasticizer loadings that reduce the polymer content below 35% may be considered excessive. Also, a coarse grind of fillers makes the sealant more susceptible to degradation by water. All finished formulations are tested for adhesion to three standard surfaces—glass, aluminum, and concrete—both before and after water immersion. Thus, the various fillers and plasticizers are evaluated for their retention of physical properties and adhesion characteristics.

Adhesion additives have to be used to obtain adhesion of the sealant to a number of surfaces. Silanes are used predominantly in sealants used for insulating glass, since silanes are good for glass adhesion and metal adhesion. Certain phenolic resins have been found satisfactory for adhesion to aircraft metals, but blocked phenolic resins were needed to prevent staining of masonry when used in building sealants.

Coal tar is used as a plasticizer for highway sealants, but its only benefit is to lower cost. A combination of coal tar and coal tar distillates is the best, but keeping the volatile content low requires careful formulation. Polysulfide coal tar compounds were used for highways, aircraft runways, and canal sealing, and a number of federal specifications were derived to cover various applications. In recent years, the technology has been updated and hot-melt PVC compounds are now widely used with new ASTM specifications covering the various application areas.

The LP-3 and LP-33 polymers of lower molecular weight and higher activity have been used to react with epoxy resins. The reaction products had lower hardness, were no longer brittle, and found use in electrical potting, as adhesives for new concrete to old, as adhesives for terrazzo overlays, and as concrete coatings for bridge decks and garage ramps.

8.4. Properties

Specific physical and chemical properties are needed for different areas of application. Sealants based on polysulfide liquid polymer have peculiar properties inherent to the polymer. This will be discussed in areas of specific interest to users.

8.4.1. Odor

The basic polysulfide polymers have a peculiar but not disagreeable odor that even carries into the cured sealants. The odor is not harmful since the volatile components are extremely low, but it is distinctive. Attempts to mask the odor by the introduction of more volatile masking agents and perfumes was found more objectional than the basic odor and was not pursued. Cured sealants can be identified by odor, and if this is difficult, then burning a small piece of cure sealant will give off a very distinctive odor specific to polysulfides.

8.4.2. Toxicity

Extensive studies were run on all the basic polysulfide polymers for both toxicity and allergy. None of the standard polymers are toxic when taken orally, and do not cause any allergic skin reaction. However, the lower molecular weight polymer LP-3 and LP-33 are skin soluble, and therefore care should be used in handling these polymers. Sealant can be removed from the hands by solvent, which unfortunately degreases the skin to ultimately cause cracking. Some of the compounding ingredients can be harmful. Lead dioxide, which is toxic, can lead to lead poisoning if ingested. Some of the one-component sealants have used barium oxide as a dehydrating agent during manufacture, but barium is toxic and cartridges containing barium have to be plainly marked as poison. Some of the phenolic resins in themselves can be allergenic if not handled properly during sealant manufacture.

8.4.3.   Solvent Resistance

The base polymers when cured using only good reinforcing fillers display good resistance to many solvents. This is one reason for their use as integral fuel tank sealants. With proper plasticizer, the sealants are still unaffected by moisture. Some plasticizers and fillers and other compounding ingredients can be hydrophilic, and this is one reason for running screening tests on all components in deriving a finished formulation. When the sealants are to be used in a chemical environment, specific formulations have given satisfactory performance as on jet fuel aprons, laboratory floors, and chemical laboratories. Federal specifications were developed for sealants used on airport pavements, where jet fuel was a problem.

8.4.4.   Adhesion

Although adhesion additives are used to develop adhesion of sealant to some standard surfaces such as aluminum and glass, primers are needed for other surfaces including masonry. In the early days, commercially available primers were tried and recommended for special surfaces such as plastics, stainless steel, zinc, copper and various masonry surfaces, but this was not the best way in the long run. The final solution was to develop a number of primers to be used against various surfaces along with standard formulations, and make this information available to sealant manufacturers. They used this basic information with their own sealant formulations and made changes accordingly. As a result, all sealant manufacturers have their own primers that have been developed for their own sealants. Primers are not interchangeable from sealant to sealant.

The formulation of primers is an art more than a science, since the end result in the use of any primer cannot be predicted. Nevertheless, certain basic data has been established; but in the end the Edisonian approach is needed to finalize the best primer formulations. Although there are a number of A-stage phenolic resins, only a select few with the use of compatible plasticizers have been found satisfactory for polysulfide sealants. Certain polyester resins with select plasticizers have also been found successful for use on certain surfaces. Phenolic resins are good for masonry and porous surfaces. Numerous silanes are available with many reactive terminals including hydroxyl, epoxy, amine, and even carboxyl. Selected silanes have been found very specific for some metal, plastic, and glass surfaces. Silane used as an adhesion additive can be different from that used in a primer formulation. The high reactivity of silanes with polar groups require that they be kept dry during storage, and be used at very low concentrations to essentially lay down a monomolecular film on metals, glass, and plastics. Primers for masonry and porous surfaces require a much higher solids content to seal off the porosity of the substrate and keep water away from the interface.

Although only several classes of active primer ingredients have been mentioned, there are many other chemical compounds that have been found for very specific surfaces or that work with heat activation. The sealant formulator is always looking for simple and more universal formulations, but until that day arrives, there will be many primers for specific surfaces, for each class of sealant.

8.4.5. Hardness

The hardness of rubber materials is measured by a durometer such as a Shore A hardness gauge. The values are relative and only have meaning when the concept is completely understood. An uncured sealant has a value of 0. A poorly cured sealant or a highly plasticized sealant with no filler would have a value of 0 to 5. It is also necessary to know whether the hardness reading is instantaneous or is given after 1, 2, or even 5 seconds after contact. A polysulfide sealant might have a reading 10 to 15 points lower after 1 or 2 seconds delay, since polysulfide sealants display fast stress relaxation. The drop-off would be lower for silicone sealants, since they have very high recovery; and would differ again for other classes of sealants. Solvent-release sealants would have an even greater drop-off after any time delay. Certian ASTM specifications and test methods differ in the time delay, and the architect should definitely be aware of these differences. Sealant manufacturer's data sheets may not differentiate the time delays, since it is to the sealant manufacturer's advantage to show lower values, particularly after heat aging.

A properly cured polysulfide building sealant would have an instantaneous Shore A hardness of 15 to 30. Any sealant that had reached a hardness of over 45 might have too high a modulus to give high movement capability. Coal tar modified sealants for airport runways would fall in the range of 5 to 10. Sealants for use on sidewalks would require a hardness of approximately 25 or more to withstand the penetration by women's spike heels. Sealants are affected by cold temperatures and eventually become rock hard when their freeze temperature is reached. Polysulfide sealants do become stiffer at lower temperatures, but their hardnesses are not greatly affected at temperatures down to $-30°F$. At this temperature the sealants are still flexible, but would exhibit a 10 to 20 point rise in hardness. The most important area, which is generally overlooked, is hardness with time or after heat aging. Properly formulated sealants should not increase in hardness over 45 on an instantaneous reading after 6 weeks' heat aging at 158°F. This is the most important hardness value for any polysulfide sealant.

8.4.6. Aging and Weathering

Good polysulfide building sealants have good resistance to aging and weathering. They do not have as good weathering characteristics as the silicones, but will successfully stand up after an exposure of more than 1000 hours in the weatherometer; they may therefore be considered very good for exterior use. Long exposure to UV and sunlight will cause some crazing on the surface, but this does not extend much below the surface with properly formulated sealants. However, the UV is much higher in the tropics, and some manufacturers have not recommended polysulfide for this environment.

Properly formulated sealants should not exhibit much increase in hardness after long exposure outdoors. Sealants made with excessively volatile plasticizers, on the other hand, can shrink, craze, and crack with time. If the volatile loss is high, then the hardness will also increase greatly, causing eventual adhesive or cohesive loss with subsequent movement. Laboratory heat aging is needed to differentiate between good and poor sealants, and most standards are not severe enough to effect a good separation. The recommendation of a maximum of 45 Shore A hardness after 6 weeks at 158°F using the instantaneous hardness value would separate the good from the questionable sealants.

8.4.7.   *Ultraviolet Resistance*

Polysulfide sealants have fair to good resistance to UV exposure. Some sealants for buildings may show light surface crazing after UV exposure in laboratory testing, but in temperate climates these sealants can perform satisfactorily if properly formulated. Some sealants have been used in the glazing area with satisfactory performance; Lever House is a glazing application that has been satisfactorily sealed for over 25 years. The cure mechanism used for insulating glass is based on manganese dioxide, which gives better resistance to UV and better retention of adhesion to glass after UV exposure through glass. Units made using polysulfide sealant will pass ASTM E-773 standard. These sealants are not as good as silicone or urethane in UV resistance, and where such exposure is expected in greater concentration, care should be used in selection of sealant.

8.4.8.   *Physical Properties*

The testing of experimental sealant formulations in most cases follows standard rubber-testing technology. The sealant is either cast or molded into a thin sheet and dumbbell specimens died out and tested on standard equipment at a standard rate of separation of 20 inches per minute. The values are converted into tensile data, giving psi and percent elongation as well as various modulus values. The average architect reading physical properties of sealants does not realize that the values are based on testing a cross-section of the dumbbell specimen 0.080 inches in thickness, 0.25 inches in width, and 1 inch long. This thin band can give very high elongations, and very misleading tensile values. Such values only have meaning

TABLE 8.4. Tensile Adhesion Values on an ASTM Test Specimen

Elongation (%)	Tensile Adhesion (psi)	Heat-Aged Tensile Adhesion (psi)
10	8	15
20	13	26
30	16	33
40	16.5	37
50	18	40
75	20	47
100	20.2	52

when compared to values for similar compounds, such as the results of a study shown in Table 8.4. All sealants must be cured exactly the same, and tested in the same identical manner. Tensile data derived from dumbbell has no meaning when attempting to describe the performance of a building sealant. Unfortunately, much sealant data is based on dumbbell data, and everyone looks for high tensile strength and high elongation but has no criteria for making any conclusions.

The physical properties of sealants would best be based on the ASTM test specimen, which is ½ × ½ × 2 inches in cross-section with tensile values given for 25, 50, and 100% extension after original and heat-aged cure of 6 weeks at 158°F. The rate of separation should be at a speed of 0.2 inches per minute. Table 8.4 shows the type of values that would be obtained on a good building sealant.

The heat aged samples are placed in a 158°F oven for 6 weeks before testing. Test specimens can extend several hundred percent, but values beyond 100 percent are, again, meaningless. This testing is done on a cross-section of sealant that more resembles a normal joint configuration. The values around 25% have the most meaning since this is the maximum movement expected of the sealant in the joint. The rate of movement in Table 8.4 was at 0.2 inches per minute rate of separation. This is satisfactory for comparative test data but too fast for the tensile adhesion studies in ASTM C-920, which only requires 1/8 inch per hour. Each type of test is significant for the purpose intended. The rate of separation of joints in buildings can be very slow if the joint is uninhibited, or very fast if the joint has resistance to free movement. The British Research Station has measured aluminum substrates and found movements up to a rate of 120 inches per hour or 2 inches per minute. Very rapid movement can cause spalling of concrete, since the tensile strength of concrete is only about 300 psi. A tough sealant with a sudden jerk could have values on tensile adhesion of 300 psi or greater on an instantaneous basis. Panek has found that tensile strength is proportional to the logarithm of the speed of separation. In other words, if the tensile adhesion value at 25% was 50 psi when measured at a rate of separation of 2 inches per minute, the tensile value could increase to over 300 psi at about 65 inches per minute, which is not unreasonable with slip/stick movement on curtain walls. The tensile adhesion values should be within the 10 to 20 psi range at a 0.2 inches per minute rate of separation, so that the 50 psi value is very high for normal application. Generally normal movement is slow and the greater problem is stress relaxation if the joint is kept at an open position for too long a period of time.

Another problem that can occur in the normal joint is that when the tensile adhesion value becomes too high, the value could exceed either the adhesion value to the substrate or the tensile strength of the sealant. In either case, the joint will fail, and the sealant will either fail in adhesion or cohesion. It becomes readily apparent that the tensile adhesion values should be on the low side, and that testing will ultimately determine whether the sealant can be used in a joint with low or high movement.

8.4.9. Creep and Stress Relaxation

These two properties define the amount of internal flow that takes place in the sealant. Polysulfide sealants occupy an intermediate place on the scale, between the very high recovery of silicones and the very low recovery of the solvent-release acrylics. A polysulfide sealant when extended 50% willl flow internally and relieve the stress by as much as 50% in the first 30 minutes. This property can be both an advantage and a disadvantage, but it is necessarily accompanied by a corresponding lack of recovery. If the sealant is kept in an extended position for a long time, the sealant configuration becomes more elongated. If kept in a closed position, the bulge can become permanent and with subsequent extension, can cause an irregular joint configuration, as discussed in Chapter 4. The amount of movement capability can be determined by testing against Class 25 or Class 12.5 in the ASTM C-920 standard. If the sealant passes the requirements of Class 25 it is capable of ±25% joint movement. If it fails Class 25 but passes Class 12.5, then it is capable of at least ±12.5% joint movement. Tests against both classes include a compression cycle that is quite rigorous, and if the sealant is used within the limitations established in the standard it should perform satisfactorily in the field.

8.4.10. Application

The two-component polysulfide arrives at the job site in a proportioned kit. The curing agent is mixed with the base material immediately prior to use. A pot life of 3 hours is normal at a temperature of approximately 75°F. The pot life can become one hour or less at 100°F, and applicators should be aware of this possibility. Sealants for use on horizontal pedestrian areas can be accelerated to cure faster. For vertical building joints, the material is usually hand mixed by using a paddle and a slow-speed mixer at approximately 200 rpm. Adequate mixing can be accomplished in 3 to 5 minutes, and is easy to recognize if the two components are of a different color—streaking indicates incomplete mixing.

The polysulfide sealants are made in a non-sag consistency for vertical building joints and in a pourable consistency for horizontal joints. The sealants extrude well from any standard caulking gun. They are sufficiently flowable to fill any voids or irregularities in the substrate. Joints can be tooled with a pointing tool or a small spatula dipped in solvent. The polysulfides are very thick at temperatures below 60°F and should be conditioned at approximately 75°F for at least 24 hours to facilitate mixing and transfer into cartridge and gunning. Polysulfides adher well to steel, aluminum, and glass, generally without a primer, and primer is recommended for porous substrates such as wood and masonry. Special cure mechanisms give the sealant good performance on insulating glass, which is still a large volume outlet for polysulfide sealants. Pricing is competitive with urethane and solvent acrylic sealants and slightly lower than silicone sealants. Because of fierce competition, the quality of polysulfide sealants has declined, and Thiokol withdrew from its "seal of approval" program in the late 1970's because sales were insufficient to support this expensive testing program. While good polysulfide sealants can still perform, these sealants will never again enjoy a healthy market. Tables 8.5 and 8.6 compare the advantages and disadvantages of one component and two-component sealants.

8.4.11.   *Reference Standards*

Good two-component polysulfide sealants can meet federal specification TT-S-00227E and ASTM C-920. Reference standards for one- and two-component polysulfides have been adopted in Australia, Canada, England, Germany, Japan, and other countries. One-component polysulfide sealants

TABLE 8.5.   Advantages and Disadvantages of One-Component Polysulfides

Advantages	Disadvantages
1. One-component sealant	1. Requires moderate temperature for faster cure
2. Broad color range	2. Requires high humidity for faster cure
3. Good durability	3. Slow cure at low temperatures
4. Good adhesion	4. Poorer recovery
5. Can meet TT-S-00230C	5. Limited package stability
6. Can meet ASTM C-920	6. Not recommended for pedestrian traffic areas
	7. Not recommended for sidewalks
	8. Slight odor

TABLE 8.6. Advantages and Disadvantages of Two-Component Polysulfides

Advantages	Disadvantages
1. Overall better physical characteristics: recovery, adhesion-in-peel, and tensile-adhesion	1. Requires mixing, but easily mixed
2. Fast through cure	2. Slower cure below 40°F
3. Better UV resistance	3. Light colors a problem
4. Better water resistance	4. Limited pot life
5. Life expectancy over 20 years	5. Very short pot life at 100°F
6. Non-staining to masonry	6. Slight odor
7. Can meet TT-S-00227E	7. Poorer UV resistance compared to urethane and silicone sealants
8. Can meet ASTM C-920	8. Poorer recovery compared to urethane and silicone sealants
9. Cost slightly lower than one-component since tubing and labor are expensive.	9. Primers needed for porous substrates

have difficulty meeting federal specification TT-S-00230C, but over the years a selected few have performed satisfactorily. One-part specifications have been approved in Canada for slightly lower requirements as well as in England. The ASTM C-920 standard is written for both one- and multicomponent sealants, and although many companies claim to meet both standards, the user should have this certified by an approved testing laboratory. Thiokol had its own certification programs during their approval program, and their standards resembled the federal specifications but were easier to meet.

Additional recommendations for upgrading ASTM C-920 are given in Appendix 2. These upgraded accelerated tests and requirements are recommended for polysulfides, urethanes, and silicones that are expected to perform for 20 years with no failure.

Over the years, a number of federal specifications were written around polysulfide sealants. At least 25 federal standards were written covering aircraft adhesion, coating, and sealant systems. Today probably the only active specifications relating to polysulfide sealants are the integral fuel tank sealant specifications MIL-S-7502 and MIL-S-9902. This application area is still carefully controlled and only certified manufacturers can supply sealant. This market is enjoyed by a few West-Coast companies that have kept close contact with the needs of this industry and have an untarnished record of quality production.

Another area enjoyed by polysulfide sealants at one time were the highway, airport runway, and canal sealants. Standards were written by various state and federal agencies covering requirements in these areas. Today, these standards have been replaced by ASTM standards, which can be met with hot-melt formulations and will be discussed in Chapter 19.

8.5. Summary

The polysulfide sealants began as one of the first elastomeric sealants for the modern curtain wall. They enjoyed good success for a number of years until competition caused a lowering of quality. This factor, plus the introduction of urethanes and silicones that have better UV and ozone resistance along with other good properties, began the competitive era. Good polysulfide sealants can still perform in a number of areas, but they only enjoy a minor part of the building industry in the U.S. today. Their recovery is poorer than that of urethanes and silicones, which is a factor in meeting performance specifications. Polysulfides still have a major part of the insulating glass market, but here again there are better systems available that do, however, require special production equipment, special handling, and are generally much higher priced.

9

Silicone Sealants

9.1. History and Chemistry

The earliest work on silicones probably dates to the synthesis of silicontetrachloride by Swedish chemist Johan Berzelius in the early 1800s. Boot (13) gives some early history on silicone chemistry in various parts of the world. In the 1930s the General Electric Company and the Corning Glass Works began work on developing high-temperature electrical insulations. Corning Glass and the Dow Chemical Company formed the Dow Corning Corporation to carry out their developments, and used the Grignard process for their early production of silicones. GE looked for less expensive methods and developed a process in which silicon metal is reacted directly with an alkyl halide such as methyl chloride to give chlorosilane intermediates. These products can then be reacted with water to form basic polymers, which can then be converted to reactive polymers. The typical reactions are as follows:

$$
\begin{array}{c}
\quad\quad CH_3 \quad\quad\quad\quad\quad CH_3 \quad\quad CH_3 \\
\quad\quad\; | \quad\quad\quad\quad\quad\quad | \quad\quad\quad\; | \\
Cl\!-\!Si\!-\!Cl + H_2O \rightarrow HO\!-\!Si\!-\!\Big[O\!-\!Si\!-\!\Big]\!-\!OH \\
\quad\quad\; | \quad\quad\quad\quad\quad\quad | \quad\quad\quad\; | \\
\quad\quad CH_3 \quad\quad\quad\quad\quad CH_3 \quad\quad CH_3 \; _n
\end{array}
$$

Intermediate A

$$
\begin{array}{c}
O \\
\| \\
CH_3\!-\!Si(OCCH_3)_3 + \text{Intermediate A} \rightarrow
\end{array}
$$

Reactant A

$$\left[\begin{array}{c} CH_3 \\ | \\ HO-Si- \\ | \\ CH_3 \end{array} \left[\begin{array}{c} CH_3 \\ | \\ O-Si- \\ | \\ CH_3 \end{array} \right]_n \begin{array}{cc} CH_3 & O \\ | & \| \\ O-Si-(OCCH_3)_2 \\ \\ \end{array} \right] + \text{ acetic acid}$$

<u>Intermediate B</u>

Intermediate A is dimethyl polysiloxane and can be either formulated or reacted to produce polymers that will cure at room temperature. Reactant A is triacetoxy methyl silane and introduces very active acetoxy terminal groups which in the presence of moisture coreact to give the cured polymer and acetic acid. This is the representative of the earlier silicone sealants that give off acetic acid upon curing. Other reactants can be substituted for Reactant A to give other terminals such as amine, amide, phenol, alkoxy, to give a wide range of polymers that have various activities, and even include polymers that can be painted over.

Intermediate B is the base polymer that reacts with moisture in the air to give the cured polymer and acetic acid. Other base polymers can be made by combining suitable reactants with Intermediate A.

There are now four major manufacturers of silicone sealants, General Electric, Dow-Corning, Staufer-Wacker, and Rhone-Poulenc. Rhone-Poulenc in France is affiliated with Rhodia in the States. Union Carbide is involved in the manufacture of silicone intermediates, and Mobay Chemical Corporation has recently been involved in selling reactive silicone intermediates to licensees who can formulate sealants. At the present time the four major suppliers sell a finished sealant, and this has stabilized the industry since all suppliers furnish sealants with high-quality performance. In addition, a number of sealant manufacturers have become distributers for several of the four major manufacturers. There are several companies that manufacture quality silicone sealants through a license arrangement with Mobay. These companies also supply sealants to other distributors. Companies that now distribute silicone sealants include Tremco, Pecora, Perennator, Polymeric Systems, Gibson-Homans, DAP, C. L. Laurence, and others. Silicones are now enjoying rapid growth and some companies are developing hybrids; unless the quality is protected, the industry could become fragmented.

9.2. Compounding

The basic silicone polymer is clear and colorless, which permits a wide latitude in compounding. A clear sealant can be made using fumed silica, and is used in some areas where some degree of transparency is needed. The clear sealant is limited in its outdoor use since UV can affect the adhesive bond. For outdoor use, opaque sealants are needed to withstand UV and a wide selection of colors is possible. There are at present five standard colors, but this can change at any point. Fillers that can be used include silica, calcium carbonate, titanium dioxide, and clay.

Only silicone oils can be used as plasticizers, since all organic ester-type plasticizers are insoluble in the polymer. These silicone oils are liquid silicone polymers of various molecular weight with nonreactive terminals. The filler content for structural sealants is generally kept at less than 15%, but can be increased to contents up to 60% where high physical properties and higher hardness are required, such as in certain gasket applications. The modulus of the sealant is controlled by the amount of nonreactive polymer or oil used in formulating the sealant. Sealant formulations are generally simpler, but nevertheless, catalysts, adhesion additives, cross-linking agents, selected fillers, and other ingredients are needed in most formulations. Silicone sealants are generally made with little or no solvent, which means low shrinkage and no hardness increase or modulus increase. The one-component sealants are unusual in that even though there may be some moisture infiltration such as in a drum, the exposed sealant may cure to a rubbery layer, which when removed exposes unaffected sealant. Silicones are unusual in that they are less affected in viscosity by temperature changes than any other polymer. This is an advantage in cold-weather sealing, since the sealants do not have to be kept at any minimum temperature for ease in extrudability.

9.3. Properties

The outstanding property of silicones is that the sealants have very high recovery. Specimens compressed and held for one year may show as much as 98% recovery upon removal of the load. Even the low-modulus sealants exhibit recovery in the order of 70 to 90%. With this type of recovery, it is no wonder that the low-modulus sealants are being recommended for ±50% joint movement applications. The standard sealants are recommended for approximately ±25% joint movement, and sealants are now being developed for intermediate ranges.

The pigmented silicone sealants are unaffected by UV and ozone, and once cured exhibit very little hardness increase with time. Selected sealants can be made to adhere to most sufaces, either by themselves or with the use of specific primers. It is best to contact the specific sealant supplier, since the supplier's recommendations could contradict another supplier. The anodized coating on aluminum may be a problem with some sealants, and it is given special consideration when stopless glazing is involved.

General Electric has developed a very comprehensive chart for three of its sealants, and has three specific primers that can be used along with specific cleaning instructions for obtaining adhesion to over 40 surfaces. The surfaces include masonry, various types of glass, metals, plastics, rubbers, and painted rather than unfinished wood. Dow-Corning has also done exceptional work in new application areas such as highway joints, and has solutions for most standard surfaces. Some of the other sealant manufacturers selling silicone sealants, such as Tremco, have also done excellent groundwork.

The picture on silicone sealants has changed drastically in the last five years and is expected to change even more in the near future. Thus far, there has been no dilution of the final product, but only time will tell.

In the uncured state, silicone sealants have an excellent gunning consistency and very good shelf stability, which make them very popular with the craftsman on the job. An opened cartridge of one-component sealant need not be used all at once. The operator has to force a little bit of sealant into the plastic tip of the cartridge and then let it cure. This plug of cured sealant effectively closes the opening for approximately one month. To reuse the cartridge this small plug of cured material can be pulled out and removed and work can proceed.

Another factor that makes the silicones popular on the job is their exceptional stable viscosity. Temperature has little effect on the gunning characteristics of the sealant. The silicone sealants can be easily extruded from the caulking gun over a temperature range of $-40°F$ to over $200°F$. Of course, cold weather will retard the cure.

The curing of silicones is now quite variable. Some one-component sealants give off acetic acid, which can be slightly irritating. Adequate ventilation will eliminate most problems here. Other one-component silicones can release an amine, a phenol, or an alcohol. These systems are recommended for various surfaces and also used to give lower-modulus sealants. As with any chemical, precautions are given for working with these systems. Two-component silicones are now available for fast cure in the plant for the manufacture of insulating glass units where a quick cure is required and also large joints are used. The acetic acid sealant can also be used here if desired. The use of silicone sealant for insulating glass gives excellent adhesion to glass which is unaffected by UV, but must be used with a polyisobutylene sealant to reduce moisture vapor transmission (MVT). The use of silicone sealant alone will result in water condensation inside the unit in a relatively short period of time, because the MVT of silicones is much higher than polysulfide and would permit water vapor to be absorbed into the unit and eventually exhaust the desiccant used in the spacer. The use of a thin bead of polyisobutylene as a primary seal (see Figure 9.1) reduces the MVT to a very low number—thus correcting the one deficiency of silicone. The combination of silicone/butyl gives the best performance in insulating glass units on the market today. Table 9.1 compares the MVT values of various sealant used in manufacturing insulating glass.

The results in Table 9.1 were obtained using a Phthwing cup and ASTM procedure E-96. Cast sheets were made approximately 80 ml in thickness

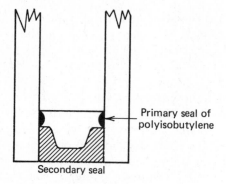

Primary seal of
polyisobutylene

Secondary seal

FIGURE 9.1. Dual seal for insulating glass. Primary seal is a bead having approximately 5 mils thickness. Secondary seal function is to hold unit together.

and water placed in the cups. The cups were tightly closed, and then placed in a drum packed with Molecular Sieve desiccant at a temperature of 75°F. The cups were weighed daily for 2 weeks for the higher MVT sealants, and for 3 months for polyisobutylene, and the values plotted against time. The slope of the straight line was used to determine the MVT. There are other methods and other parameters used to measure MVT, but careful study must be made of conditions before making any comparisons of MVT. There is no fast method for determining MVT, since water can be trapped in the filler of the sealant or released by the cure mechanism, and several weeks under the test environment above are needed for the system to get into equilibrium.

The use of polyisobutylene with silicone, polysulfide, and even hot-melt butyl lowers the MVT of the unit to that of the polyisobutylene—thus considerably stabilizing the dew point change with time. Since satisfactory dew point after testing is a criterion of satisfactory performance, the use of polyisobutylene becomes apparent.

In the cured state, silicone sealants give very high recovery. Shore hardness will seldom vary on a temperature range of −40 to +180°F. The medium and high-modulus sealants have a Shore hardness of about 30 to 35. The low-modulus sealants have hardnesses of approximately 15. High-modulus silicones will give at least 95% recovery after extended periods of time up to 1 year, which accounts for their use in structural or stopless

TABLE 9.1.   MVT Values of Various Sealants

Polymer System	MVT in g/m²/24 hours
Silicone	16–24
Polysulfide	6–16
Hot-melt butyl	1–4
Urethane	6–16
Polyisobutylene	0.1–0.2

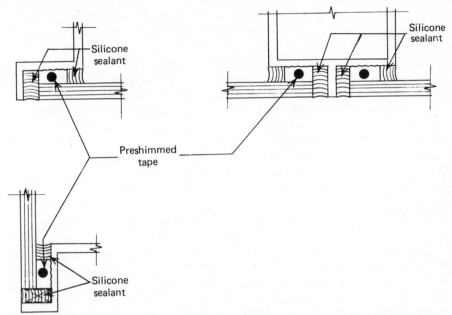

FIGURE 9.2. Stopless glazing system. One of various systems designed to give two areas of sealant contact. Five-foot-square lites of glass tested at Construction Research Laboratories finally broke upon attaining 210 psf, equivalent to 350 mph. (Courtesy Tremco, Inc.)

glazing (see Figure 9.2) Low-modulus silicones also exhibit good recovery, on the order of 60 to 80%, and can also meet ASTM C-920 for Class 25 and claim performance for ±50% joint movement. The silicone manufacturers are trying to get adoption of a Class 50 in the ASTM C-920 standard or a separate specification. Some silicone manufacturers are making claims for −100% extension and 50% compression, but this may be very optimistic. This claim is for highway joint sealants.

High-modulus silicones have poorer tear resistance, and since the sealant is always under strain when extended, tears will propagate if the silicone is near its limit of extension. Karpati (14) has indicated that high-modulus silicones can perform more satisfactorily if the movement limits are restricted to ±22%. The sensitivity of high-modulus silicones to tear has undoubtedly resulted in the introduction of the low-modulus silicones, which have greater movement capability. These sealants are being recommended in higher movement areas including glazing but not stopless glazing, and for building joints and even experimental highway joints. Rather than attempting to give numbers, polysulfide sealants will show better tear resistance than the high-modulus silicone sealants, but the low-modulus silicones will be better than polysulfide.

Any attempt to give numbers to ultimate tensile strength and ultimate elongation is meaningless, since the manufacturers now make materials that would encompass a very wide range in these properties. The only logical parameters are performance against ASTM C-920 in addition to more stringent requirements for maximum hardness, which should be held to 40 after 6 weeks at 158°F. The maximum weight loss should be held to 6% under the same conditions. All test assemblies should also be tested under the accelerated conditions. The low-modulus silicones should not be expected to increase beyond 25 Shore A hardness or to lose more than 10% in weight under the accelerated conditions. If claims are made for higher movement, such as ±50%, then all tensile adhesion cycling should be done at these extensions and all tests run before and after accelerated heat aging.

9.4. Available Materials

A number of silicone sealants and manufacturers have been cited. In order to clarify the various sealants, Table 9.2 lists the current suppliers, types of sealants, and other pertinent data. Table 9.3 lists advantages and disadvantages of silicone sealants as compared to other classes of sealants.

The high-modulus acetoxy polymers are recommended for stopless glazing, glazing, and most building joints with the possible exception of masonry. Low-modulus sealants are recommended for glazing and most building joints, and some sealants are specifically modified for use on ma-

TABLE 9.2. Various Silicone Sealants

Manufacturer	Product	Modulus	Terminal	±Movement Capability (%)
General Electric	1200	high	acetoxy	25
	2400 (Silglaze)	low	acetoxy	50
	Silproof	low	alkoxy	50
	1700	high	acetoxy	25
Dow-Corning	790	low	amine	50
	999	high	acetoxy	25
	795	low	? (alkoxy)	50
	888	low	? (alkoxy)	over 50
Tremco	Proglaze	high	acetoxy	25
Rhodia	Rhodotherm	high	acetoxy	25
	3B	high	acetoxy	25
SWS	951	high	acetoxy	25
	930	low	amine	50
	940	low	amine	50
Mobay	3135	high	acetoxy	25
	3196	low	amide	25
	3123	low	amine	25

TABLE 9.3. Advantages and Disadvantages of Silicone Sealants

Advantages	Disadvantages
1. One-component sealant	1. Slightly more expensive
2. Colors available	2. Limited color range
3. Color stable	3. Critical surface preparation
4. High temperature resistance	4. Possible concrete adhesion
5. Low temperature gunability	problems
6. Excellent UV resistance	5. Dirt pick-up
7. Excellent ozone resistance	6. Poor tear resistance with high
8. Nonstaining	modulus
9. High movement capability	7. Slight odor problem
10. Very high recovery	8. Aluminum surfaces a problem
11. No shrinkage	with some sealants
12. No hardness increase with time	9. Short tooling time
13. Improved tear resistance with	10. Primer selection needed for
low-modulus sealants	various surfaces
14. Various moduli available	
15. Medium to high movement	
capability from ±25 to ±50%	
16. Long durability over 20 years	

sonry. Generally these have other terminals than acetoxy. GE 1700 is specifically formulated for sanitary areas such as bathrooms. Dow-Corning 888 is recommended for highway joints. For insulating glass the following sealants can be used: GE 1200, Dow 999, Tremco Proglaze, and Rhodia Rhodotherm. The Mobay basic polymers can be modified to fit all areas above. Sealant manufacturers must be contacted for specific instructions for surface preparation and the use of primers.

The pertinent properties for building sealants are movement capability, hardness, hardness increase with time, weight loss with time or after heat aging, adhesion, and recovery. Parameters such as ultimate tensile strength and elongation, with modulus have little meaning. The classification for movement capability either directly or indirectly ties in with all properties.

TABLE 9.4. Recommended Ranges of Properties for Three Movement Classes

Property	+25%	+40%	+50%
Original Shore A hardness	20–30	10–20	5–15
Heat-aged hardness	35 max.	25 max.	20 max.
Maximum % weight loss	6	8	10
Adhesion-in-peel pli	5	5	5
Recovery	95%	80%	70%

FIGURE 9.3. United Nations Plaza Apartments. Sealed with a one-component silicone sealant in 1965. (Courtesy of General Electric Company, Silicone Division.)

Table 9.4 illustrates recommended ranges of properties for three movement classes. Sealants meeting these requirements in tests against ASTM C-920 using 6 weeks at 158°F as the heat aging cycle should perform satisfactorily for over 20 years.

9.5. Reference Standards

The federal specifications TT-S-227E and TT-S-230C originally covered all classes of sealants including silicone. However, because the original high-modulus silicone sealants had problems with some sections of both standards, a third federal standard, TT-S-1543B, was issued specifically for silicones with a slightly different test assembly.

Today with the lower-modulus silicone sealants, there are silicone sealants that will meet all the above specifications and in addition the latest ASTM standard C-920. Since the above reference standards do not exceed movement capability beyond ±25%, there has been considerable activity in ASTM Committee C-24 to consider a specification for ±50% movement capability. Additional work is now being developed to support a separate specification for structural or stopless glazing.

Silicone sealants easily meet all existing reference standards, and attempts should be made to upgrade existing standards with additional accelerated conditions given in Appendix 2.

9.6. Summary

Silicones are now available in various movement capabilities from ±25% to over ±50%. They are now used in building joints including structural or stopless glazing because of the very high recovery of silicone sealants. They are quite stable in both uncured and cured state. They have excellent workability, excellent color stability, and outstanding resistance to UV, sunlight, and ozone. Table 9.3 summarizes the advantages and disadvantages of silicone sealants as might pertain to both architects and applicators. Figure 9.3 shows a picture of a building sealed with one-component silicone in 1965 and still satisfactory.

10

Urethane
Sealants

10.1. Introduction

Urethanes were first developed in Germany and England in the early 1940s and were used to prepare synthetic rubbers. With the shortage of natural rubber during World War II, their wide versatility enabled the urethanes to be developed in a number of application areas. It became apparent that the synthetic products offered greater control and greater latitude and resulted in many new products.

The development of urethanes was gradual, but in the early 1970s they were being used in considerable quantity. Urethane sealants were introduced in the early 1970s, but only as two- or multipart sealants and then later as one-part. Urethane foams became a major item. Urethane paints with excellent abrasion resistance were developed for gymnasium floors. Extremely tough plastics having very high abrasion resistance were developed for use on factory vehicles, skate boards, and other areas requiring tough abrasion resistance. Other products included fibers and thermoplastics.

The basic reactive group in forming any urethane polymer is an isocyanate group which, when reacted with a hydroxyl group, yields a urethane group as shown in the following equation:

$$R—N—C—O \; + \; R^1—OH \rightarrow R—O—\overset{\displaystyle O}{\overset{\displaystyle \|}{C}}—\overset{\displaystyle H}{\overset{\displaystyle |}{C}}—N—R^1$$

$$\text{isocyanate} \qquad \text{alcohol} \qquad\qquad \text{urethane}$$

The groups R and R' can contain additional reactive groups, so that if both reacting groups are difunctional a polymer chain can be built up.

Both water and amines can also react with isocyanate groups as follows;

130

$$H_2O + OCNRNCO \rightarrow H_2NRNH_2 + CO_2\uparrow$$

$$\begin{array}{ccccccc} & H & & H & O & H & & H \\ & | & & | & \| & | & & | \\ H_2NRNH_2 + OCNR^1NCO \rightarrow & -N-R-N-C-N-R^1-N-C- \end{array}$$

urea group

The first reaction forms an amine plus CO_2 gas that is the basis for making foams. The second reaction results in the formation of a urea group in the repeating polymer chain. These essentially are the basic reactions to form all types of urethane products. However, the selection of base components determines the type of product desired. All polymer reactions require a diisocyanate and a number are available. For products where slight discoloration is not a problem, toluene diisocyanate (TDI) is used. Where paint products or sealants must not discolor, then hexamethylene diisocyanate (HDI) is used, which is more expensive.

For sealant polymers, long chains containing hydroxyl groups are needed. Typical reactants can include hydroxyl terminated polyester, castor oil and its derivatives, and various glycols including ethylene, propylene, butylene, and hexylene. Also triols are needed to give three-dimensional structure to the sealants, and include trimethylolpropane, glycerol, hexantriol, and others. Lastly, amines can be used either in small quantity to give non-sag properties, or as a coreactant to give tough fibers such as nylon. From here on in, the urethane polymer formulation belongs to the chemist and is guided by the costing department. Polymers with good and bad properties can be prepared by juggling the various components. Polymers with good water resistance require raw materials without too many ester groups, or polyethylene oxide groups which are cheaper raw materials. Damusis (15) and Evans (16) go into excellent chemical detail on the possible chemical reactions and raw materials that can be used.

Because of the ease of preparing urethane polymers for use in sealants, many companies have made their own base polymers, thus replacing the middleman. Other companies have bought prepared polymers which could be further modified with fillers and compounding ingredients to form one- or multicomponent sealants. Tremco coreacted a urethane polymer to give a polymer with epoxy terminals. This is then cured with a polymeric diamine giving a polymer with more work life, better UV and ozone resistance, better color stability, and higher recovery and movement capability. Products Research, on the other hand, coreacted a urethane polymer to give mercaptan terminals for quicker cure as an insulating glass sealant with improved performance over a standard polysulfide polymer. It is easy to derive polymer variations with other terminals; the problem is to derive a useful polymer that is easy to manufacture at a reasonable cost.

The two- or multicomponent urethane sealants came first. Part A contained the polymer with reactive isocyanate terminals along with the necessary fillers, adhesion additives, and plasticizers. Part B was a reactive polyol that reacted with the isocyanate terminals to give the desired sealant. Part C, when used, contained the colors that permitted both standard colors as well as the possibility of matching the color of various substrates. W' 'n Part C is not used, then the standard color components must be incorporated in Part B. The one-component sealants are tricky, since additional mechanisms have to be employed to prevent the formation of CO_2 gas, which is a normal by-product between isocyanate and moisture. One mechanism derived by Damusis and utilized by Bayer reacted the polymer with phenol, thus blocking the polymer, and then added a diketimine in the absence of moisture. When the formulated one-component sealant was exposed to moisture, the moisture-unstable diketimine was converted to a diamine, and a solvent, methyl ethyl ketone. The very active diamine split off the phenol and reacted with the released isocyanate groups in a chain reaction to give the cured polymer. Some people have classified the urethanes as intermediate in performance between the polysulfides and silicones. However, they do offer unique physical properties and consequently cannot be replaced in many areas of application. They are considered to be a high-recovery sealant with good workability, good adhesion, and good movement capability, and consequently are classified as better than polysulfides in these and other properties.

10.2. Compounding

The typical two-component sealant consists of a base polymer with isocyanate groups and a second component containing the curative, polyol. There are exceptions, such as the Tremco three-component system. The multicomponent package offers greater versatility in compounding and color matching and for a number of years was always the better sealant. The one-component sealant—contrary to general expectations—is usually more expensive, since it is more difficult to prepare, involves more reaction steps, and may involve additional catalysts. The packaging of one-component sealants requires moisture-free sealant, and additional steps in dehydrating the compound and components through all the stages in manufacture. The cartridges must be special, since they must protect the sealant from any moisture for at least 6 months from the time of delivery. One way to get long package stability is by using a longer polymer—which means fewer reactive terminals—which also results in much slower cure. This dodge was used in the early days to get around the problem of package stability. The real problem is to get a fast cure upon exposure, along with good package stability. However, there are a few sealant manufacturers that can meet this requirement and also meet ASTM C-920 for Class 25.

There are many plasticizers that can be used as well as many fillers. Cheaper plasticizers become apparent by poorer aging properties, such as higher weight loss after heat aging, and poorer adhesion retention upon water immersion. Cheaper fillers, which are coarser, such as a coarse grind of marble dust, become apparent by poorer physical properties, lower modulus, lower recovery, and higher water sensitivity. All deficiencies only become apparent through testing and performance, which is the reason for recommending certification from approved testing laboratories. The better plasticizers have lower volatility. The better grades of filler can be much finer ground or even include some precipitated filler.

A good urethane sealant would have the following general formula:

Polymer	35–45%
Fillers	30–40%
Colorants	2–3%
Thixotropic agents	1–2%
Adhesion additives	1–3%
Plasticizers	15–25%
Solvent	0–4%

The total percentage of weight loss after heat aging should not exceed 6%. Selected adhesion additives are needed to get adhesion to the surfaces desired, and the general use of silanes either as additives or primers is now quite widespread. Some sealant manufacturers still caution against the use of urethane against glass, which is a problem surface. Urethanes perform best against masonry and metals, but even here it may be best to run a mock-up trial, since all precast concrete is contaminated with form oils and primer may be necessary.

Small quantities of antioxidant and UV absorbers may be needed depending on the components used in preparing the polymer. Colorants with sealants always are a consideration, since some pigments have some activity. The use of a third component permits wider color matching with no stability problem.

10.3. Properties

Urethane sealants come in a wide range of properties some of which are pertinent to performance in building sealants. Emphasis will be placed on the important areas of consideration.

10.3.1. Hardness

Cured urethane building sealants should fall in the range of 20 to 30 Shore A instantaneous hardness. More important, the hardness after heat aging or with extended time should not exceed 40. Hardness change with time or heat aging is a major clue to poor formulation, since hardness is also associated in a deterioration of all the pertinent properties associated with joint movement, such as joint movement capability, adhesion, and surface cracking and crazing.

Hardness for traffic-bearing sealants might be increased to 25 to 35 on original cure, but again, should not exceed 40 to 45 on aging. There are sealants that can give this performance and meet ASTM C-920 requirements. If the sealant is to be used in low temperatures down to $-20°F$ it should not increase more than 5 to 10 points at the low temperatures. Failure to meet this means improper polymer selection or formulation.

10.3.2.   Odor and Toxicity

Urethane sealants have essentially no odor unless some solvent is used, in which case the solvent odor is the problem. In closed areas, ventilation should be used. Toxicity of urethanes is no longer a problem, although good housekeeping is essential to prevent additional problems of clean-up since solvent is needed. Most urethanes carry precautionary warnings since trace quantities of isocyanate might be present and might cause some allergic reaction if the sealant is not handled with proper respect.

10.3.3.   Solvent Resistance

Urethane sealants can have good oil and solvent resistance, and if used in a chemical environment, the manufacturer's literature should be checked for specific instructions. Urethanes can withstand rain, but are no longer recommended for continual immersion as in swimming pools, since the adhesive bond eventually breaks down.

10.3.4.   Aging and Weathering

Good urethane sealants have excellent resistance to UV and ozone, and will not crack or craze even after long exposure in the laboratory or outdoors. Good sealants will exhibit good recovery and retention of recovery with time and also good retention of movement capability with time. Some light-colored urethanes might discolor with time, but if color retention is a requisite, then pretesting under UV laboratory exposure should be carried out.

10.3.5.   Tensile Strength and Elongation

Ultimate tensile strength and ultimate elongation have no meaning when tests are carried out using ASTM D-412 as the test method. Using dumbbell specimens gives elongation values ranging from 300 to 1000%, which have no correlation with joint movements that are not expected to increase beyond 25%. The measurement of tensile properties using ASTM D-412 only have importance when used by the chemist in making comparative studies. The important property for sealant is tension adhesion measured on the ½ × ½ × 2-inch specimen cited in ASTM C-719 and discussed in the next section.

10.3.6. Tension Adhesion

The only logical dimension for testing a joint sealant is one that resembles an average joint, namely the ½ × ½ × 2-inch specimen between parallel plates used in ASTM C-719 for measuring tension adhesion. Although this test assembly can be stretched beyond 100%, it also gives very practical values in the range of 25 to 100% for joint movement. An average elastomeric joint sealant should be expected to give a value of approximately 10 psi (the assembly has a cross-section of 1 square inch) at 25% elongation. The value can increase up to 20 psi at 100%. The assemblies can be heat aged, and such assemblies should not increase in value by more than 50%. Testing heat-aged assemblies is an excellent check on the stability of the sealant. The cohesion–adhesion test cycle is run on this test configuration and is the most severe part of ASTM C-920. The assembly should be used in describing the physical properties of all elastomeric sealants.

10.3.7. Recovery

Good urethane sealants will exhibit recovery on the order of 70 to 90%, which puts them in a much better class than polysulfides, but not quite as good as silicones. The high recovery of urethane accounts for the good performance in building sealants, and the ability to meet sealant specifications for both one- and two-component sealants.

10.4. Application

Multicomponent urethane sealants are packaged in proportioned kits for delivery to the job site. The separate components are usually viscous liquids which upon mixing become thixotropic for use in vertical and horizontal joints. As with polysulfides, the base components should be stored at approximately 70°F for ease in mixing. Since the components are dissimilar in color, poor mixing is apparent by the visual marble effect, which disappears when the sealant is thoroughly mixed. Most multicomponent sealants will give 5 to 8 hours of work life, which is much longer than for polysulfides and not as temperature-sensitive. Complete cure is obtained in 24 to 36 hours under normal conditions. The sealants will be quite firm in less than 24 hours.

The one-component sealants are usually supplied in ⅒-gallon cartridges and ready for use. The cartridges have a minimum of 6 months' package

TABLE 10.1. Advantages and Disadvantages of Urethane Sealants

Advantages	Disadvantages
1. Can be used in joints up to 6 inches	1. Light colors can discolor
2. ±25% movement capability	2. Poor water immersion resistance
3. Excellent recovery	3. Not recommended for wet joints
4. Excellent UV resistance	4. May require more priming
5. Excellent ozone resistance	5. Limited package stability for one-component
6. Fast cure for multicomponent	6. One-component requires more cure time
7. Long work life for multicomponents	7. Multicomponent requires mixing
8. Negligible shrinkage	8. One-component not recommended for traffic areas
9. Excellent tear resistance	9. Not recommended for stopless glazing
10. Excellent chemical resistance	
11. Excellent durability (20 to 30 years)	
12. Can meet ASTM C-920 for all systems	
13. Much better than polysulfides	

stability from the time of delivery, will skin cure in a matter of hours, and can cure in 1 to 3 weeks under normal conditions.

Most urethanes require priming to various surfaces, but claims vary among the manufacturers. Also, there is a difference of opinion with regard to the use of urethanes for glazing. The general concensus is that urethanes are not recommended for glass. Urethane sealants are now available for use in making insulating glass units which should meet existing standards.

Urethanes are priced in the same general range as silicones and may be slightly higher than polysulfides. Price no longer dictates the selection of sealant. Urethanes do an outstanding job on building joints, and are the choice for pedestrial and traffic areas. With their better UV and ozone resistance, they are more widely used than any other class of sealant for buildings, with silicone sealants in second place and polysulfide falling off fast in this area of application. Table 10.1 compares the advantages and disadvantages of urethane sealants.

10.5. Movement Capability

One- and two-component urethane sealants are available that have a movement capability of ±25%. These sealants can be used for building and traffic joints. There was a time when the traffic joints were limited to ±12.5% movement, but this has changed within the past two years. Some epoxy-urethane sealants are said to have a +40% to −25% movement capability.

Because of the activity with low-modulus silicone sealants to meet ±50% movement, and the consideration of a standard for this class by ASTM Committee C-24, some companies have begun testing modified urethane sealants to the ±50% class. As of November, 1983, at least one manufacturer has indicated a possibility of meeting the requirements for this class, so that at least another polymeric group is going to challenge the low-modules silicone sealants in the ±50% class.

10.6. Reference Standards

Both one- and multicomponent urethanes will meet ASTM C-920 and the two outdated federal specifications TT-S-00230C and TT-S-00227E. The urethanes will also meet Canadian specifications 19GP15 and 19GP16. All the above reference standards are borderline with respect to potential quality, and good sealants should be able to meet the recommended amended standard in Appendix 2, which should be used by specifiers desiring better urethane sealants.

10.7. Summary

Urethanes have become the main sealant for building joints, with silicones second and polysulfide now a poor third. The transition was partly due to the competitive picture that lowered polysulfide quality. Urethane sealants become prominent because of greater flexibility in polymers and better performance. The chemical input resulted in many types of sealant systems, which eventually outperformed polysulfide in many areas. While urethanes are slightly higher priced, architects have recognized the better performance, and price is not a prime factor in sealant selection. Urethanes have a place in the future of building sealants—a place that will be challenged by silicones.

11

Solvent-Based
Acrylic Sealants

11.1. Introduction

Solvent-based acrylic sealants fall into a semi-elastomeric class of sealants since they do exhibit thermoplastic properties. When solvent-based acrylics were introduced in the late 1960s they were recommended in areas beyond their capability, but because they performed very well when applied properly, they found their proper niche in the sealant market. These sealants perform best in the area from ± 7.5 to $\pm 12.5\%$ movement capability. Their movement is limited because they harden with time due to the loss of solvent used in the system. Their main advantage is that they adhere to a wide variety of surfaces without the need for a primer. The use volume is presently greater than for polysulfides but less that that for silicones or urethanes. These sealants have a place in the market because they are one-component, have a self-healing feature, and have excellent adhesion.

11.2. Compounding

The basic acrylic building block is acrylic acid $CH_2\!\!=\!\!CH\!\!-\!\!COOH$. The two methods of altering the starting raw material is by converting the acid to an ester and by substituting alkyl groups for the alpha hydrogen. This gives rise to a wide family of acrylic monomers that, when polymerized, yield a variety of polymers. Many of the polymers are based on mixtures that yield more useful properties with regard to sealant applications. Some of the starting materials used to make solvent acrylic sealants are shown in Table 11.1.

TABLE 11.1. Reactive Acrylic Monomers

Acrylic acid	$CH_2 = CHCOOH$
Methyl acrylate	$CH_2 = CHCOOCH_3$
Methyl methyacrylate	$CH_2 = CCOOCH_3$
	$\quad\quad\quad\quad\; \mid$
	$\quad\quad\quad\quad CH_3$
Ethyl acrylate	$CH_2 = CHCOOC_2H_5$
Methacrylic acid	$CH_2 = CCOOH$
	$\quad\quad\quad\quad\; \mid$
	$\quad\quad\quad\quad CH_3$

Although acrylic chemistry started in 1843, the commercial production of polymers did not start until the late 1920s. The first acrylic sealant was developed in 1958 by Tremco. The manufacturers of raw materials include Rohm and Hass, American Cyanamid, and Dupont, and several sealant manufacturers are making their own polymers. From the chemical standpoint, there is an advantage to using two or more monomers in the preparation of the polymer in order to achieve the desirable properties such as elasticity, adhesion, UV resistance, hardness, and recovery. A good typical acrylic sealant would have the following general formula:

Acrylic polymer	35–40%
Fillers	40–45%
Catalysts	1%
Thixotropic agents	2–3%
Plasticizers	1–5%
Solvent	10–15%

The polymer is usually supplied as a 85% solution in xylene. The main fillers are usually ground calcium carbonate with fine-particle silica to give non-sag properties. Other fillers are needed for color, including titanium dioxide, which is necessary in all of the light-colored sealants. Small quantities of plasticizer may be needed, but this is generally limited to a small amount since its presence could lead to sag. Claims are made for a slow cure with time and some catalysts may be used for this purpose. For the most part, cure is obtained by the slow evaporation of solvent. Several grades of sealant are available, and the need for heating chests implies a sealant with lower solvent content and also less shrinkage upon cure. The percentage of weight loss after cure or heat aging is the most critical value of these sealants. The lower the weight loss the better the sealant. It must be apparent that values of 10% loss in weight must result in approximately 20% volume shrinkage, which can affect the geometry and performance of the sealant.

11.3. Properties

Solvent-based acrylics are generally nontoxic and nonallergenic. Since the sealants are solvent-based, they should be used with a normal amount of care. A solvent is needed for installer's clean-up.

Odor is a distinct problem, and should be kept in mind when working in closed areas. The trace acrylic odor is offensive, and adequate ventilation is needed. When caulking indoors, the building should be unoccupied. Also, the odor can be absorbed by food. The sealant becomes tack-free in 1 to 2 days and should not be used in pedestrian areas or on walkways, since the sealant remains soft for long periods of time. A partial cure will be obtained in 2 to 5 months, but final cure may require 1 to 2 years. This may be deceiving, since some manufacturers give a 1 to 2-year guarantee on performance of the sealant, which may be the time in which only a partial cure has been obtained.

Complete cure is accompanied by a large increase in hardness. The hardness after some cure may result in a sealant with 25 to 35 Shore hardness, but increases slowly with time. A good solvent-based sealant should not exceed a 55 Shore A instantaneous hardness after long time exposure or after heat aging. This higher hardness value is the clue in illustrating the limited movement capability of these sealants. High hardness is associated with high modulus and less movement capability. When excessive movement occurs, the sealant must either fail in adhesion or cohesion. Because the sealant behaves as a thermoplastic, it can neck down with slow movement, and is thus able to absorb all of the movement with the flow. If the movement is excessive, then the sealant will either fold over or tear apart. The best solution is to recognize the movement capability and use the sealant within its limits. Some manufacturers cite ±7.5% and others give ±12.5% as their limit.

Because of the solvent nature and slow cure, these sealants wet all surfaces, and consequently no primer or surface preparation are recommended. This one aspect of good adhesion is the main qualification for the solvent-based sealants. Adhesion values of 10 to 20 pli can be obtained to most surfaces. A good sealant has good resistance to UV and ozone, and exhibits good color stability, but light shades may show dirt.

The thermoplastic nature of these sealants results in no recovery in a standard test, which is expected with these sealants.

The sealants will perform outdoors in normal exposures. They are not recommended in any areas where any amount of water immersion is found, since some sealants will revert on contiuous exposure to water. The reversion is evident by a permanent softening of the sealant, and some flow or sag will result. Problem areas are joints at the base of parapet walls, where the wall may become saturated with moisture. This is not an application area for these sealants, and urethane sealants should be used here with proper weeping.

11.4. Application

The excellent adhesion of these sealant makes them one of the most versatile sealants for building construction. Along with the fact that only a minimum amount of cleaning is needed, no real surface preparation is necessary. They can be used for glazing, including toe and heel beads, brick to brick, brick to metal, and for most perimeter joints around doors and windows. They have been used on wood sash and wood-to-masonry joints. They can

FIGURE 11.1. Toronto City Hall. Sealed with a solvent-based acrylic sealant in 1964 and still doing fine. (Courtesy of Tremco, Inc.)

TABLE 11.2.   **Advantages and Disadvantages of Solvent-Based Acrylics**

Advantages	Disadvantages
1. Require no priming	1. Poor low-temperature elasticity
2. Require minimum surface preparation	2. Cannot be used in joints over ¾ inch wide
3. Excellent adhesion	3. Poor recovery
4. Good color stability	4. Slow skinning and cure
5. Cures tough	5. Remains tacky for several days and picks up dirt
6. One-component system	6. Limited flexibility
7. Excellent UV resistance	7. Strong, offensive odor until skins
8. Self-healing	8. Poor water resistance
9. Good durability of over 20 years	9. Not recommended for water immersion
10. Good chemical resistance	10. High cold flow
11. Nonstaining	11. Not recommended for traffic joints
12. Some sealants can take ±12.5% movement	12. Some systems only have ±7.5% movement capability

be used on control joints, precast joints, panel-to-panel joints, bedding of mullion and frames, cap beads over tape, and back bedding. While most of the above areas will have limited movement, the success of these sealants depends on not abusing the sealant with extensive movement. A building successfully sealed with a solvent-based acrylic sealant some two decades ago is shown in Figure 11.1.

Where the earlier sealants required a warm-up chest, these sealants gave better performance because they contained lower solvent content and consequently gave less shrinkage and less hardening. The market has forced the development of sealants that did not require a warm-up, with some slight loss of performance. The warm-up sealants are available with some suppliers upon special request. The advantages and disadvantages of solvent-based acrylic sealants are listed in Table 11.2

11.5.   Reference Standards

There is no ASTM standard for this class of sealants, although test methods have been derived and approved. It will be some time before ASTM Committee C-24 will derive a standard for this class.

Federal specification TT-S-00230a has been cited in part as a reference standard for these sealants. The first standard was issued as TT-S-00230 in February 1964. It was replaced by TT-S-00-230a in September 1967, with TT-S-60-230b in September 1968, and TT-S-00-230c in September

1970. The use of federal standards always implies that the latest version is the one to be used, but reference has been made to TT-S-00230 and TT-S-00230a. This is misleading since the performance requirements are much easier and these earlier specifications are almost impossible to obtain. The TT-S-00230 specifications are written around sealants having either ±12.5% or ±25% movement capability, so that the supplier's literature should be consistent with these claims.

However, the sealant manufacturer is in a difficult position, since no U.S. standard exists for these sealants, and the federal specifications were written for elastomeric and not solvent-based acrylic sealants.

There are two solutions to the dilemma. One is to cite the Canadian Standard 19GP5M, which was written around sealants having only ±7.5% movement capability but is generally a fairly good specification. The second is to use the recommended specification limits given in Appendix 3, which are based on observed and expected performance capability for good solvent-based acrylic sealants.

11.6. Summary

Solvent-based acrylic sealants are very widely used because they are a one-component sealant and will adhere to almost any surface without a primer. They skin fast, but require 2 to 5 months for a fair cure and may take up to 2 years for a complete cure. They cure to a relatively tough sealant with some shrinkage, so are not recommended for more than ±12.5% movement capability. No reference standard has been adopted in the U.S., but all the pertinent test methods have been adopted, and recommendations for a proposed specification are given in Appendix 3. The volume of these sealants in use approaches that of the urethane sealants.

12

Butyl
Caulks

12.1. Introduction

The growth of butyl rubber began in the late 1930s and early 1940s. The scarcity of natural rubber led to the construction of butyl rubber manufacturing plants by the U.S. government. After the World War II the facilities were purchased by private industry, and the basic technology was established. The first butyl caulks appeared in the construction market in the 1950s.

Butyl caulks are widely used, but the main volume is in the over-the-counter market. Some butyl is used in factory glazing, and some goes into the construction market. A considerable volume goes into tapes both cured and uncured, which will be covered in Chapter 16. Another segment goes into hot-melt applications mostly for insulating glass.

Because of its low price and excellent stability, butyl-related products are widely used as rubbers, caulks, and adhesives. Their use in caulks also covers a wide range of price and quality depending on the market area and the application.

Essentially, most butyl caulks are either non-curing or slow curing, and consequently are not recommended for large-joint-movement applications. The maximum joint movement capability is ±7.5%. The butyl caulks do not harden or craze and are quite stable to the elements. They can be made as solvent-based caulks using the tough rubber as a starting material, or can be plasticized with polybutene oils that permanently soften the caulk and do not volatilize. Consequently, the field is wide open to formulators. Although there are only a few manufacturers of butyl rubber, such as Exxon, Cities Service, Petrotex, and Polysar, there are dozens of formulators who purchase the raw materials and sell the products as specialty items including gutter sealants, driveway crack sealer, concrete block sealer, and foundation crack sealer, all of which can do a reasonable job if no movement takes place.

The greatest volume of butyl caulks is to the over-the-counter market, with only about 20% going into controlled areas of application and construction. The total volume of butyl caulks exceeds that of urethane or silicone. Irrespective of past claims of growth for these caulks, the major market is already fixed in the over-the-counter segment, and the application of butyl caulks in the construction market will be limited by their small joint movement capability.

12.2. Compounding

The chemistry of butyl polymers involves three closely related monomers as shown below:

$$
\begin{array}{ccc}
\text{CH}_3 & \text{H}_3\text{C} \quad \text{CH}_3 & \text{H} \quad \text{H} \quad \text{CH}_3 \\
| & | \quad\quad | & | \quad\quad | \quad\quad | \\
\text{HC} = \text{CH} & \text{HC} = \text{CH} & \text{H}_2\text{C} = \text{C} = \text{CH}_2 \\
| & & \\
\text{CH}_3 & & \\
\text{isobutylene} & \text{butylene} & \text{isoprene}
\end{array}
$$

The polymerization of isobutylene gives the widest range of polymers from tough rubbers to soft mastic, none of which has any potential to cure up since there are no unsaturated sites along the polymer chain. Nevertheless, these polyisobutylene polymers have extended use in caulks, adhesives, tapes, and even as modifiers in rubber compounding. The introduction of small quantities of isoprene up to 2% produces butyl rubbers that can be formulated and vulcanized to form rubber products. These rubbers are also used in some caulking formulations where some degree of structure or cure is desired. Butylene only polymerizes in part to yield oils of various molecular weight, which are used as plasticizers or solvents for the butyl rubbers and polyisobutylene. It is possible that all three polymers are used together in various caulks, tapes, and cured rubbers, since each supplies one or more desirable properties.

Butyl caulks can be made using butyl crude rubber, which can be dissolved in a solvent to give a solvent-based butyl sealant. It is also possible to use the butylene oils to soften down the butyl rubber to give nonshrinking caulks. Polyisobutylene can also be plasticized with the butylene oils to give mastics compounds that do not cure for possible use in curtain-wall joints that interlock and where the caulk must remain permanently pastic.

TABLE 12.1.   Butyl-Rubber-Based Caulking Compound

Ingredient	Percent by Weight
Polyisobutylene	2.5
Butyl rubber, 50% solution in mineral spirits	20.5
Talc powder	30
Calcium carbonate filler	20
TiO_2 pigment	3
Adhesion resin	4
Polybutene	10
Plasticizer	2
Thixotropic agents	2
Drying catalysts	0.05
Mineral spirits	6

Solids content 84% by weight

A good butyl-rubber-based caulking compound might have the general formulation shown in Table 12.1. The formulation in Table 12.1 uses butyl rubber for structure, but also uses polyisobutylene and adhesive resin for tack. Polybutene plasticizer is used to soften the compound. Some trace driers are used to give a tack-free surface and slowly cure the compound. The total butyl composition in the caulk is approximately 22%, but the total solvent is 16%, which could result in approximately 30% volume shrinkage. Most butyl caulks cure by solvent release. Some structure is desired by the use of the crude rubber which must be solvated by the mineral spirits. Additional solvent may be used to incorporate more filler into the compound. Since the over-the-counter market is highly competitive, it is not expected that high-quality caulks will be obtained in this area, and lower polymer content would be the end result.

12.3.   Properties

The butyl caulks are one-component gun-grade compounds. Tack-free time could be quite long, ranging from 6 hours to several days, and complete cure in 1 to 6 months when properly formulated. The butyl caulks are relatively soft, low-modulus materials with little or no recovery. They have good shelf life and are available in white, bronze aluminum/stone, and black for the industrial market, and gray for the home market. The properties of only the industrial-grade caulks will be discussed, since these are the only compounds made to existing standards. Although the over-the-counter market is considerably larger, these caulks may at best have minimal requirements to meet, and no attempt will be made to cover their performance properties.

12.3.1. Hardness

Butyls are relatively soft materials that will set to approximately 5 Shore A hardness within 1 to 2 weeks. Materials that will cure will reach 20 to 30 hardness in several months. Because of the availability of the polybutenes it would be most unusual to see butyl caulks with hardness over 30 with aging.

12.3.2. Physical Properties

Because of the slow cure and the general low modulus of these sealants, it is impossible to express any so-called tensile strength, elongation, or even tension-adhesion values for any butyl caulks. Most butyl sealant suppliers do not even give a maximum hardness, although federal specification TT-S-001657 gives a value of 40 after heat aging. The testing of butyl caulks is based on mastic type materials rather than elastomeric tests. These is no ASTM reference standard for butyls, but there are two standards for oil-and-resin base caulks and these test methods are pertinent to butyl caulks. The movement capability is limited to $\pm 7.5\%$ and the maximum hardness should not exceed 40 after 3 weeks' aging at 158°F. The adhesion-in-peel should be a minimum of 4 pounds per linear inch, and the adhesion after UV exposure should not be affected. The maximum weight loss should not exceed 25%, and there should be no failure on test assemblies after exposure to UV and $-10°F$, flexing at $-10°F$, or any bubbling. The bond-cohesion test should be run with $\pm 7.5\%$ movement and there should be no failure. These values have been incorporated into a recommended standard for butyl caulks given in Appendix 4.

12.3.3. Toxicity

The butyls have no toxicity or allergenic problems. They do require the use of a solvent for clean-up. Odor is also no problem except for the odor of mineral spirits, which is the dominant solvent in these caulks. The usual precautions regarding ventilation and skin care apply to the butyls and the solvent mineral spirits.

12.3.4. Aging and Weathering

Butyl caulks age and weather with no cracking or surface crazing. They have good UV and ozone resistance and consequently can be used in exterior glazing. A well-compounded butyl will withstand 1000 hours in an accelerated weathering chamber. Higgins (17) has stated that butyl compounds become softer if any oxidation takes place in the polymer chain, since oxidation tends to break down the polymer, as compared to usual oxidation which cross-links across double bonds. This tends to further insure against any increase in hardness or modulus with time.

12.3.5. Solvent and Chemical Resistance

When cured, butyls have good resistance to acids and alkalis and good water resistance. They also have fair resistance to some oils and solvents, but the suppliers' literature should be checked for specific chemicals—urethanes might prove better in a chemical environment.

12.3.6. Recovery

The butyls are a low-recovery caulk. Creep and stress relaxation are poorer than with a soft polysulfide. The situation is further aggravated by the fact that the caulks are slow curing and can contain polybutenes and polyiso-butylenes which internally plasticize the butyls. Because of the low recovery, the caulks behave like a soft acrylic, and will conform into the new shape with any movement. Consequently, butyl caulks have a limited joint movement of $\pm 7.5\%$, which makes them very useful for domestic housing, but limited to essentially nonmoving joints in industry.

12.4. Application

Butyl caulks may be delivered to the job site in $\frac{1}{10}$-gallon cartridges or in 1 or 5-gallon pails. For the home user they are available in cartridge form. The butyls may be safely applied in a temperature range of 40 to 120°F. Below 40°F they are difficult to extrude.

Butyls require some joint cleaning, but extensive surface preparation is not required. Laitance and loose concrete should be removed from the faces of concrete joints, but priming is unnecessary. Metal surfaces may be wiped with an oil-free cloth wet with solvent to remove any surface contamination. Wood surfaces generally require wiping with a clean cloth or soft brush. Glass should be wiped clean with a dry soft cloth. Butyl caulks work best in relatively nonmoving joints. They are used on the mating surfaces of heating and air-conditioning ducts and for sealing openings where pipes and ducts pass through roofs and partitions. Butyl caulks can be used below grade on masonry joints. They may be used in the lap joints of sheet siding, metal roofs, glazing, wooden door and window perimeters, caulking of interior and exterior sills, and threshholds.

The home owner uses butyl caulks for glazing and for window and door perimeters, flashing, roofing repairs, and general crack sealing. Butyls are used by installers of storm doors and windows and protective siding. Butyls adhere to aluminum, steel, and vinyl siding.

Some butyl caulk is supplied in tape form for metal building joints where a controlled thickness of sealant bead is highly desired during fabrication.

One specific area for good butyl caulks is as a caulk over neoprene glazing gaskets that have taken a partial set and leak during a driving rain. Butyls

TABLE 12.2. Advantages and Disadvantages of Butyl Caulks

Advantages	Disadvantages
1. Reasonable cost	1. Very slow cure
2. Availability	2. High shrinkage
3. Good flexibility	3. High compression set
4. Good adhesion to most substrates	4. Limited to joints with ±7.5% joint movement
5. One-component	
6. Little surface preparation	5. Not recommended for expansion joints
7. Good water resistance	
8. Good color stability	
9. Four colors available	
10. Only material for capping neoprene gaskets	

are the only sealant that will adhere to neoprene gaskets, but have a limited service life of 3 to 5 years since the adhesion is gradually fugitive due to plasticizer content in the gasket. Table 12.2 compares the advantages and disadvantages of butyl caulks.

12.5. Reference Standards

There is no ASTM standard for butyl caulks, although test methods covering the properties have been approved. A specific recommended standard is given in Appendix 4.

There is a federal specification, TT-S-001657, which was written for butyl caulks, covers some of the pertinent properties, and would be satisfactory for use as a stopgap specification. This specification is written for ±7.5% joint-movement capability.

While federal specification TT-S-00230 may be cited, there are no butyl caulks in industry that claim ±12.5% joint movement, which is one of the two classes covered by 230. Consequently, any claims to meeting this specification should be backed up with certification from an approved testing laboratory.

A Canadian standard covering butyl caulks with ±5% joint movement is 19GP14. This is a fairly good butyl standard and covers better-quality caulks than those covered by TT-S-001657.

12.6. Summary

Butyl caulks are single-component sealants that set to a rubber-like consistency. They generally are 70 to 80% solids, which is accompanied by shrinkage in the joint. They are a low-modulus caulk with low recovery.

The industrial grade gives good performance when used properly. The grade for the home owner is probably satisfactory for nonmoving joints.

Butyls do not compete with polysulfides, urethanes, or silicones except where their lower requirements for movement are matched. Butyls compete with latex caulks, solvent-based sealants, and oil-and-resin-based caulks in both the construction and consumer markets, although each has optimum areas of application. Butyl caulks have the advantage of low cost, which may be the deciding factor where materials of equal performance have been found.

Butyl caulks appear to be recognized for their limited performance capabilities and their limitations, particularly in the construction market, and the relative volume will remain static over the next several years.

13

Latex
Sealants

13.1. Introduction

The growth of latex sealants paralleled the growth of latex paints. As recently as 1956, oil-based paints were the only available materials for exterior use. By 1966, latex paints had taken over 40% of the exterior paint market and 80% of the interior market. Today, the values are approximately 75% for exterior and 80 to 90% for interior. The development of latex sealants was a natural outgrowth of the latex paints.

The chief materials for latex sealants are acrylic, vinyl acrylic, and polyvinyl acetate. The use of acrylic latex gives a more rubbery product; the use of vinyl acrylic gives a slightly less rubbery product and is also cheaper; and polyvinyl acetate gives the stiffest sealant at still less cost and is more sensitive to UV. Acrylic latex sealants can be used outdoors in limited application areas with a maximum joint movement of $\pm 7.5\%$. Sealants based on vinyl acrylic or polyvinyl acetate are best used indoors, where temperature gradients are smaller and no water resistance or UV resistance is required. Latex sealants are all one-component gun-grade compounds. They have fair flexibility, but little recovery, and because of their water-base nature they have high shrinkage—on the order of 30% by volume, which would give 15 to 20% weight loss.

Latex sealants offer the advantage of very easy clean-up with a damp cloth, and only water is needed to clean hands and tools. Latex sealants are used mostly in light construction, such as residences, schools, churches, small industrial buildings, and apartments. They may compete somewhat with solvent-based acrylics in the outdoor market, but are dominant in most indoor applications.

Painting contractors are the prime buying influence, and most of their purchasers are through the local hardware or paint store, where consumers tend to be price-conscious. Nevertheless, with the drastic change in cost of materials in the early 1980s, there is no longer any cheap sealant on the market, and a somewhat better grade of sealant is available at the hardware store today. Items include putty, butyl sealants, solvent-based acrylics, latex sealants, neoprene caulks, and even the high-priced silicone that is highly desired for bathroom tiles and shower stalls. The 25-cent tube has been replaced by the $1.25 to $2.95 tube for any general sealant. On this basis, the smaller contractor can deal directly with the major sealant manufacturer for better-quality sealants, and sealant manufacturers have been contacting smaller contractors in recent years. Although there are many small manufacturers of latex and solvent-based sealants, most large sealant manufacturers now make these and other lower-performance sealants for themselves and others on a private-label basis.

13.2. Properties

Latex sealants are one-component gun-grade materials that are non-staining to most substrates. The are generally tack-free in 1 hour and can be painted over almost immediately. They have 80 to 85% solids content, but because of their stress relaxation the sealants usually dry into a concave shape. If the sealant has been placed in a wide joint, and some shrinkage cracks occur, this is visible in a short period of time, and the applicator can give the caulk a second pass. Figure 13.1 shows a tooled joint versus a latex sealant which gives a curved appearance after drying and shrinking.

Latex sealants have moderate modulus and low recovery. Although rubbery in appearance, they do not have high extensibility and will crack if over-extended. Most sealant suppliers put the movement capability at ±7.5%, and the materials are never put into major expansion joints. The sealants will cure to 20 to 25 Shore A instantaneous hardness, but the tear resistance is poor because of the nature of the material. The binder is latex, which attempts to fuse together on drying, but the consistency of the cured sealant is much more heterogeneous than a sealant made using a liquid polymer or a solvent solution of a polymer that thoroughly wets the filler

Latex caulk

FIGURE 13.1. Comparison of sealant beads.

and gets better reinforcement due to more intimate contact with the filler. There are no curing agents used in latex sealants, and so once the sealant has dried, it is cured. The sealants do not harden with aging since there are no volatile plasticizers, and the latex does not cross-link any further.

The sealants can be applied to damp surfaces and will adhere to most building surfaces without the use of a primer. Acrylic sealants can be used outdoors in limited areas of application and will withstand rain, but any amount of continuous water exposure will lead to loss of adhesion, since the dried sealant still contains the same emulsifying agents that were needed to keep the latex from agglomerating before formulation or in the finished sealant. For this reason latex sealants are best used indoors where water is not a factor and movements are minimal in a controlled environment. Also, the absence of UV eliminates the problem of surface discoloration. Latex manufacturers have introduced materials with higher solids content and also internally plasticized the polymer in order to get more movement capability, but any slight improvement will not be sufficient to approach the properties and performance of solvent-based caulks or the more elastomeric sealants such as polysulfides, urethanes, and silicones. Latex sealants have their fair share of the market, which is essentially the indoor market. Latex sealants have been formulated so that the cartridges will withstand several freeze-thaw cycles, which permits some laxity in storage.

A typical acrylic latex sealant formulation is shown in Table 13.1. As seen from the formulation, the compound is essentially chalk and latex. A small quantity of ethylene glycol is used as the so-called anti-freeze com-

TABLE 13.1. Formulation for a Typical Latex Sealant

Ingredients	Percent by Weight
Acrylic latex @ 55% solids	40–42
Surfactants	1–1.2
Plasticizers	7–8
Mineral spirits	1–2
Ethylene glycol	1–2
Ground calcium carbonate	46–50
Miscellaneous fillers	1–3

Miscellaneous Properties	
Sealant solids	80%
Binder	23%
Filler	45%
Shrinkage	25%
Shore A hardness	40
Adhesion-in-peel to wood	4 pli minimum
to aluminum	4 pli minimum
to glass	4 pli minimum

ponent. Small quantities of surfactants are needed to keep the sealant from coagulating in the cartridge, and a small quantity of plasticizer may be needed to make the sealant more elastic, but may be eliminated if an internally plasticized latex is used.

Properties of the sealants are based on tests designed for these materials since tests for rubbery sealants do not apply. A desirable property is no discoloration after UV exposure. Adhesion must be maintained to substrates such as wood, aluminum, and glass. Shrinkage should not exceed 30%, and the sealants must withstand several exposures to 0°F with no cracking or coagulating.

13.3. Application

Latex sealants are available in $1/10$-gallon cartridges and in 1 and 5-gallon containers. Latex sealants can be applied at temperatures above 40°F. The sealants are made to withstand some temperature below freezing. Canadian reference standard 19GP17 requires that sealants withstand 18 hours at 0°F with no freezing. The sealants are only recommended for relatively nonmoving joints. Most suppliers recommend the sealants for indoor use only where a quick-skinning and easily painted sealant is required. The sealant can be used as an acoustical sealant in dry wall construction or for plaster wall systems. Latex sealants are suitable for joints on precast concrete plank ceiling joints, indoor joints, interior door and window frames, and for tub and shower stalls. The polyvinyl-based sealants cure harder, but have more adhesive properties. They may be used to stick loose wall and ceiling tile back into place, glue wood, or for other miscellaneous adhesive applications.

Some manufacturers recommend latex sealants for relatively nonmoving

TABLE 13.2. Advantages and Disadvantages of Latex Sealants

Advantages	Disadvantages
1. Fast skinning and cure	1. High shrinkage
2. Immediate paint-over	2. Poor water-immersion resistance
3. Some flexibility	3. Poor low-temperature flexibility
4. ±7.5% movement capability	4. Can freeze below 0°F
5. No bleed through	5. Limited movement
6. One-component	6. Mostly indoor use
7. Good adhesion	7. Poor water resistance
8. Good UV resistance	
9. Easy application	
10. Easy clean-up	
11. Good package stability	
12. No primer needed	
13. Low cost	

exterior joints, but the applicator would be better off using solvent-based acrylic sealants which, although more expensive, will do a more foolproof job.

The market for latex sealants is the over-the-counter market and the nonindustrial segment for most indoor wall and ceiling work. Advantages and disadvantages of the sealants are given in Table 13.2.

13.4. Reference Standards

Canadian specification 19GP17 is a good standard covering acrylic latex sealants. This is written for ±5% joint movement. A very satisfactory standard for latex sealing compounds is ASTM C-834, and the requirements and referenced test methods are given in Appendix 5.

13.5. Summary

Latex sealants are basically designed for light construction, which is mostly residential. They have replaced to a great extent the oil-and-resin-base sealants in this market. They offer the advantages of being one-component, quick drying, and easy to clean up. They have some movement capability, can be painted over almost immediately with latex paints, and are low-cost. They have high shrinkage, poor water immersion resistance, and low recovery, but serve admirably in all indoor areas.

Although the over-the-counter market is large for latex sealants and other caulks and paints, one has only to reflect on the amount of material that is eventually discarded after the job is done. The usual home owner buys anywhere from 25 to 50% excess material for any job, which in 1 year either coagulates, thickens, skins, or discolors, and is eventually discarded. However, when the do-it-yourselfer does the job, he or she saves over 90% on the total job. On today's market, it can cost $10 or more for a gallon of paint, but it will cost 10 to 15 times as much to get a painter to apply it. So the home owner can well afford to waste some materials.

The latex sealant market is about equivalent in volume to the combined market for silicones, urethanes, and polysulfides. However, the dollar value is considerably less. The cost of a cartridge of good latex caulk is approximately $2, which is considerably less than any elastomeric caulk. Cartridges of latex sealants are also cheaper than butyl sealants, solvent-based sealants, and oil-and-resin-base sealants. One of the factors for the low cost is that there is very little energy needed to prepare a sealant, since the fillers are simply mixed into the latex. No solvent is used, which brings these sealants down to one of the lowest raw-material costs for all sealants.

The overall market for latex sealants is very bright. They are firmly established as a major component in the residential area, and will continue to sell in this area for years to come. Not much growth potential is left, however, since the market growth potential has been achieved.

14

Oil- and-Resin-Based
Caulks

14.1. Introduction

Oil-based caulks were the first sealants to be used commercially, and performed quite satisfactorily in high buildings up to the 1940s when the large skyscrapers were still made of solid masonry walls with little or no movement. With the introduction of the modern curtain wall and the new elastomeric sealants, the oil-based caulks took a back seat. The introduction of polysulfides and then the solvent acrylics, urethanes, and finally silicones, replaced a very large volume of oil-based caulks. The quality of oil-based caulks was improved with the new synthetics in an attempt to hold some of the market. The various butyls were added to give more flexibility and introduce some movement capability. In spite of improvements, however, the oil-and-resin-based caulks have now probably reached the low plateau of volume, and will probably hold their own for the next decade.

Most of the larger sealant manufacturers have discontinued their supply of oil-and-resin-based caulks, because they were not competitive with the smaller manufacturers and their cheaper caulks.

The total market is on the order of polysulfide sealants, with about 60% of the market being sold over-the-counter and a major portion of the remainder going directly to factories for factory glazing. The average home owner might buy an occasional pint can for routine glazing or crack filling, but a variety of other caulks are now available for home owners to choose from for their various problems.

The oil-based caulks are made using oils such as linseed, fish soybean, tung, castor, and others. Some of the oils may be drying oils, which quickly oxidize and harden. Today, modifiers commonly used to permanently

156

plasticize the caulks include polybutene, polyisobutylene, and even some butyl rubber. These softer caulks are now recommended in relatively non-moving joints with up to ±5% joint movement, but the hardening oil-based caulks have a maximum movement capability of ±2%.

The same principles hold true. If the caulks are not abused, they can perform over 10 years with satisfactory service. There are several good ASTM reference standards that have been recently approved and should be cited whenever good-quality caulk is desired.

14.2. Compounding

The formulation of good oil-and-resin-based caulks parallels the formulation of other caulks and sealants, in that there has to be a proper ration of filler to binder, and the sealant or caulk has to set or cure within a desirable period of time to give a product that will meet existing standards. The requirements for movement for oil-based caulks are much less and the raw material cost is lower, which is the reason why these materials find a use in their segment of the market. Factory casement windows can be satisfactorily glazed with caulks meeting ASTM C-669. Of course solvent-based acrylic or silicone sealants would do a better job, but at a higher cost, which is the reason for using these caulks in this area of application.

The desirable features for the oil-and-resin-based caulks are the fast skinning and reasonable flexibility good caulks can give. The flexibility is obtained by the use of some polybutene. Caulks for use on building joints must also exhibit fast skinning, but then remain flexible within the mass of the caulk for long periods of time. None of the reference standards give a hardness requirement.

A good-quality gun-grade oil-based caulk might have the general formulation shown in Table 14.1. A knife-grade compound would contain a less viscous oil, and would contain 20 to 25% oil. Less fibrous fillers would be used with slightly more ground carbonate filler and very little solvent.

TABLE 14.1. General Formulation for an Oil-based Caulk

Ingredients	Percent by weight
Boiled linseed oil	25–30
Ground calcium carbonate	45–50
Polybutene	5–10
Colorants	3–5
Thixotropic agents	2–4
Paint dryers	0.1–0.3
Mineral spirits	10–12

More polybutene might be added for improved knifing ability. For industrial applications, the caulks can be obtained either in cartridges or in 1, 5, or 55-gallon drums.

The quality of an oil-based caulk for home used would be considerably poorer, with probably 90% chalk and 10% binder such as boiled linseed oil, with 1 to 3% solvent. Its main function would be to quickly harden and set.

14.3. Properties

The oil-based caulks for the homeowner generally will set quick and hard with practically no movement capability, and will probably not be made to any standard. The Shore A hardness of a glazing putty could reach 70 or more in 1 year.

Materials to meet ASTM C-570 must exhibit less than 20% volume shrinkage, 180° flexibility in a set compound, good adhesion to glass, and tack-free time. Movement capability should be on the order of ±5%. Such materials would cure to a Shore A hardness of 30 within 1 year and would then gradually creep up to 50 in several years.

Materials to meet ASTM C-669 are of a higher quality. One specific point in this specification is that no dimension of any sash can exceed 20 inches. This specification is designed for face and channel glazing of metal sash. This limited size is in keeping with the limited movement capability of this caulk, which is also in the range of ±5%. The requirements of this standard expose the caulk to 100 and 300 hours in a UV weatherometer at 140°F, after which time the caulk must not exhibit more than minimal cracking or peeling, loss of adhesion, wrinkling, or oil exudation. There are no requirements of maximum weight loss, but a caulk meeting these heat-aged requirements must also exhibit some flexibility, which implies some medium hardness in the 30 to 40 Shore range, and the weight loss must be less than 20% in order to maintain some degree of flexibility.

14.4. Application

The oil-and-resin-based caulks are furnished in both knife and gun grade, and may be delivered to the job site in ¹⁄₁₀-gallon cartridges and 1 to 5-gallon pails. Hardware stores supply caulks in ¹⁄₂-pint, 1-pint, and even quart cans. The materials are easy to apply and require essentially no priming except for a quick dusting. Priming may be desirable with wood substrates, but is unnecessary with other building materials.

Applicators and contractors use the caulks for interior and exterior glazing of wood and metal sash, door and window frames, interior crack sealing,

TABLE 14.2. Advantages and Disadvantages of Oil-Based Caulks

Advantages	Disadvantages
1. Can remain plastic	1. No recovery
2. Tool easily	2. Little flexibility
3. Apply easily	3. Can harden or crack with poor quality
4. One-component	
5. No primers needed	4. Movement limited to ±2% to ±5%
6. Lowest-cost caulk	5. Not recommended for moving joints
7. Good color stability	
8. Can last over 10 years	6. Slow cure rate
9. Good package stability	7. Can stain substrates
10. Some low-shrinkage caulks	8. Few good-quality caulks
11. Good for low movement	9. Shrinkage can reach 20%
12. Fast skinning	

lap joint sealing in interior duct work, copings, and other applications. The homeowner uses these caulks for door and window framing, glazing, crack filling, and baseboard and plaster cracks. For exterior use, the homeowner would now probably use solvent-based caulks for better movement capability. Unfortunately, the homeowner may not have the best of materials available at the hardware store—but at present prices, better materials are now available. Advantages and disadvantages are given in Table 14.2.

14.5. Reference Standards

As already mentioned, the industrial market has two good ASTM standards available, namely ASTM C-570 for "Oil- and Resin-Base Caulking compound for Building Construction," and ASTM C-669, "Glazing Compounds for Back Bedding and Face Glazing of Metal Sash." These specifications cover good-quality caulk and are the best available for the next 10 years or more.

Federal specifications TT-S-001657, entitled "Interim Federal Specification: Sealing Compound, Single Component, Butyl Rubber Based, Solvent Release Type (for buildings and other types of construction)," was issued in 1970 and covers good-quality caulks. This standard essentially replaces all the outmoded standards listed below:

1. TT-C-00598c for oil-and-resin-based types.
2. TT-G-410E for back bedding and face glazing.
3. TT-P 00791b for linseed oil type.
4. TT-P-781a for putty and elastic compounds for glazing metal sash.

There are three Canadian specifications, issued in 1970 and 1971, on oil and modified oil-based caulking compounds. 19GP1b is a minimal standard for materials that are suitable for interior and exterior wood sash. 19GP2b is a standard for nonhardening modified oil-based caulk for application to wood and metal sash, and for face and channel glazing. 19GP6a is a solvent-release oil-based standard for sealing interior and exterior joints in wood, masonry, and other structures, where movement capability is on the order of ±2%. These standards are not as rigorous as the ASTM standards.

14.6. Summary

The oil-and-resin-based caulks are non-elastomeric caulks designed for joints with little or no movement. They have some shrinkage, and will gradually increase in hardness since they are based on drying oils. They are one of the lowest-priced caulks on the market and are available in various quality grades. They are easy to apply, have no handling or storage problems, do not stain, and do not require any joint cleaning or priming. Used within their small joint movement capability of ±2 to ±5%, they should perform at least 10 years.

Over the past 20 years, the oil-based caulks have lost a considerable segment of the market to more exotic types of sealants, but they will continue to be used in glazing areas and nonmoving wood and metal sash. There is no growth potential for the oil-based caulks, but the market has bottomed out and should keep a constant value for the next 10 years.

15

Specialty
Sealants

Several other sealants have captured minor portions of the market. Some have been developed because they apparently possessed unusual properties, but they have not been widely developed and are made by only a small number of manufacturers. Other sealants or caulks merely represent a variation on a theme, with no unusual properties except perhaps a slightly lower cost. Some polymers have been used to prepare sealants in the past, and the development may have been abandoned. Lastly, there are specialty polymers and sealants that have been introduced into a related area that merit some clarification. Although the introduction of a new polymer implies potential superior properties and performance, the development in the market inevitably faces costs and competition and the eventual performance proof before wide acceptance. Silicones were a long time in coming into dominance, but are now the classic materials. There is talk of hybrid silicone sealants, such as acrylic-silicones and urethane-silicones, that might result in a marriage of the good properties of the silicones and the lower costs of the hybrid polymer. As in most hybridization studies, the recessives become dominant, and it is rare that a new species is developed with no bad habits. Nevertheless, the challenge is there, and chemists are always at work developing new polymers for potential application in the sealant market. Over the past 40 years oil-based-caulks have been replaced by polysulfides, urethanes have taken over this position as the prime sealant today, and silicones are now challenging the position of urethanes.

The various specialty sealants and short-lived polymers will be discussed separately.

15.2.   *Chlorosulfonated Polyethylene Sealants*

Chlorosulfonated polyethylene (CSPE) is sold as Hypalon, which is the registered trademark of related polymers produced by Dupont. The CSPE polymer is prepared by substituting chlorine and sulfuryl chloride groups into polyethylene. The chemical modification changes the rather stiff plastic into a flexible and rubbery polymer. The sulfuryl chloride groups provide active sites for effecting cross-linking or vulcanization. The original polyethylene has a molecular weight of approximately 20,000. CSPE has a chlorine atom occurring about every 6 or 7 carbon atoms, and a sulfuryl chloride group about every 100 carbons. The CSPE elastomer is completely saturated and cross-linking can only occur at the sulfuryl chloride site. Cross-linking can occur by reaction with a polybasic metal oxide or a polybasic metal salt of a weak acid together with an organic acid and a sulfur-containing rubber type accelerator. Water is a catalyst in the process, which becomes the activator in a one-component sealant. Typical sealants would contain approximately 25% polymer, 25% plasticizer, approximately 25% filler, 10 to 15% solvent, and the rest curing agents and stabilizers. Only a few companies manufacture CSPE sealants and the market is very small compared to the total sealant market. These sealants are said to meet federal specification TT-S-00230C with a movement capability of $\pm 12.5\%$. If the claim is proper, then these sealants would also meet ASTM C-920 for Class 12.5. Essentially, these sealants could be classified as a solvent-based CSPE sealant and would meet the proposed solvent-based specification in Appendix 5.

CSPE sealants cure to a 25 Shore A hardness in 1 to 4 months, and have

TABLE 15.1.   Advantages and Disadvantagese of Hypalon Sealants

Advantages	Disadvantages
1. Impervious to water	1. Slow cure of 1 to 4 months
2. Remain flexible	2. Poor package stability
3. Fair recovery	3. Higher cost
4. Good chemical resistance	4. High shrinkage
5. Excellent UV resistance	5. Not for interiour use
6. Excellent ozone resistance	6. Not for traffic joints
7. Can meet TT-S-00230C for $\pm 12.5\%$ movement capability	7. Not for sidewalks
8. Can meet ASTM C-920 for $\pm 12.5\%$ movement capability	8. Limited to $\pm 12.5\%$ movements
9. Can meet proposed spec. for solvent-based caulks in Appendix 5.	9. Tough gunability
10. One-component	

superior resistance to UV. The areas of application include various precast concrete stone joints including expansion joints, butt joints, lap joints, corner joints, copings, capping, and floor joints. They can be used for pointing perimeter joints around doors and windows but not interior joints. CSPE compounds have excellent weathering resistance and white sheet material is used as a stable roofing material. The sealants have good chemical resistance against acids and oxidizing compounds. Because of their solvent nature, they have excellent adhesion to most surfaces without the need of a primer. The solvent results in approximately 15 to 20% volume shrinkage. Surface skinning occurs in about 4 hours, but the through cure is very slow. The total market for these sealants is small, since they actually compete with solvent-based acrylic sealants, which are supplied by many manufacturers. Though the properties may be slightly superior, since the claim is to be able to meet 230C at ±12.5% movement, there is nevertheless not that much difference since the solvent-acrylic sealants have been dominant in the solvent-based sealants for ±12.5% movement. The advantages and disadvantages are given in Table 15.1.

15.3. Polymercaptans

Ever since polysulfides were introduced, the fact was recognized that polymers with mercaptan terminals are easy to cure. Consequently, the search has always been for a polymer with mercaptan terminals with improved properties over the polysulfides. Polysulfides continue to increase in hardness with time, can show sensitivity to UV and ozone, do not have optimum adhesion to glass, and are also high-priced. Polymers different from polysulfides, but with mercaptan terminals have been termed polymercaptans. Over the years there have been several polymers introduced to the market.

15.3.1. Polymers by Diamond Alkali

A series of polymercaptan liquid polymers were introduced by the Diamond Alkali Company in the late 1960s and early 1970s. Diamond Alkali introduced basic polymers as well as finished sealants. The sealants were given good promotion, but the properties were not as good as polysulfides and were eventually withdrawn from the market. The cure mechanisms were quite similar to those for polysulfide, and the polymers also reacted with epoxy resins in the same typical formulations.

The basic polymers were prepared using a polyoxyalkylene backbone and terminated with an active mercaptan group. The chemistry was supposed to eliminate the sulfur—sulfur bonds and the carbon—sulfur bonds of the polysulfide polymers, which were the weaker links. A few of each did end up in the polymer after cure, but were statistically very few in number. The use of the polyoxyalkylene backbone, on the other hand, made the polymer more sensitive to water, and adhesion during water immersion became very poor as well as a major problem to resolve. Without good adhesion, the sealants were obviously limited in their use as a building sealant. While the other physical properties were satisfactory, the limited adhesion that was inherent in the backbone eventually doomed the polymers.

Diamond Alkali sold its patents to Thiokol. The introduction of the polymers at a lower price by Diamond Alkali and the promotion had spurred considerable initial interest, since everyone was looking for improvements and a second supplier of polysulfide-like polymers, but the new polymers lacked the major requirement of adhesion.

15.3.2.   Polymercaptans by Hooker Chemical

In 1974, Hooker Chemical introduced several polymers designated as mercaptan-terminated polyethers. These polymers were supposedly easy to make and were essentially a hydrocarbon with sulfur linkages and mercaptan terminals. These polymers had approximately 55% sulfur compared to the Thiokol polysulfide polymers, which had 37% sulfur.

The Hooker polymers had lower specific gravity and would probably have sold at a lower price than polysulfides. The chemical and physical properties of cured sealants were somewhat similar to those of cured polysulfides. Adhesion properties were poorer, but the chief complaint was the stronger odor. The cure mechanisms were somewhat similar. Except for a possibly somewhat lower cost, the polymers did not offer any improved properties and were poorer in adhesion. Several polymers were made in the pilot plant and all the major sealant manufacturers were contacted and sampled, but not enough interest was generated to warrant getting into production. The venture was abandoned after a short period.

15.3.3.   Polymercaptans by Products Research

Products Research introduced their Permopol sealants based on their own patented polyurethane polymers with mercaptan terminals. They have sold these systems as polysulfide sealants for use in the insulating glass market. The sealants have exhibited improved properties over conventional polysulfides, but have been lacking in one area. The Permopol sealants have been slower curing than polysulfides in the insulating glass market, and also exhibit long surface tack. For the insulating glass market, it is desirable that sealants are capable of being cured in 3 to 8 hours even with the help of a controlled environment such as 100°F with 75% relative humidity. Sealants not meeting these requirements would be classified as slow curing. If these problems could be solved, the Permapol sealants would be competitive with polysulfide sealants for this market. Improvements have supposedly been made, but only formulated sealants are sold. To date, no polymer has been made available, but the solution of the cure problems would go a long way in getting a larger segment of the polysulfide market. These sealants have not been prepared for building sealants since they are a two-component sealant designed for the fast cures required in the insulating glass market. Also, the insulating glass market is now the large portion of the total polysulfide sealant market.

15.4. Neoprene Sealants

Neoprene sealants were among the first elastomers offered to the construction industry. The one-component gun-grade sealants cure very slowly to a Shore A hardness of 35 to 40. They become tough, have good water immersion properties, have a movement capability on the order of ±12.5%, and have good chemical resistance. They are made with a solvent content of approximately 40%, which results in high shrinkage, and have limited adhesion to some surfaces. Their main desirable property is that they are one of the few sealants compatible with asphaltic concrete and bitumen. Consequently, they are used as crack fillers for blacktop driveways, and for cracks between concrete highways and asphalt shoulders. This outlet alone is very large when one envisons the thousands of miles of highway in this country, and is a large percentage of the volume for this sealant. Its compatibility with bitumen makes it a prime candidate for crack repair on asphalt roofs and asphalt flashing.

The sealants are also compatible with cured neoprene rubbers, and can be used as a lubricant adhesive for preformed neoprene highway joint seals. They can be used as adhesives in neoprene sheet roofing and for metal roofs. The total volume of neoprene sealants is very small, but the sealants find specific use because of their unique compatibility with asphalt, bituminous materials, and cured neoprene gaskets. The advantages and disadvantages of neoprene sealants are given in Table 15.2.

TABLE 15.2. Advantages and Disadvantages of Neoprene Sealants

Advantages	Disadvantages
1. Compatible with bitumens	1. High shrinkage, up to 40% by volume
2. Compatible with asphalts	
3. Compatible with neoprene gaskets	2. Only dark color
	3. Very slow cure
4. Low-cost	4. Not recommended for dynamic joint movement
5. Good water resistance	
6. One-component	5. No specifications
7. May have ±12.5% movement capability	6. Might meet solvent-based proposed spec. in Appendix 5 or 6
8. Good adhesion to bitumens	7. Stains stone
9. Good adhesion to asphalts	8. Stains wood
10. Good adhesion to metals	

TABLE 15.3.   Advantages and Disadvantages of SBR Sealants

Advantages	Disadvantages
1. Very low cost	1. Shrinkage up to 45%
2. Good for glazing wooden windows	2. Very poor U.V. resistance
3. Good for siding	3. Very poor ozone resistance
4. Good for hidden joints	4. Not for exposed surfaces
5. Good for gap filling in automobile manufacturing	5. Limited service life
6. One-component	6. No existing specification
7. Dries to a rubber	

15.5.   SBR Sealants

Styrene-butadiene rubber is another elastomer that has found limited use. The tough rubber requires larger amounts of solvent, and finished sealant might contain approximately 35 to 40% solvent. The sealants are low-cost with high shrinkage, and find only a limited automotive use as a gap filler, and in factories for sealing wooden windows. There is some small volume sold to the over-the-counter market as a caulk for siding and masonry joints. The total volume is small and its main advantage seems to be lower cost in limited areas of application. The sealants are made using about 15% polymer, about 20% of various plasticizers, 40% solvent, and the remaining 25% fillers. The sealants have poor weathering resistance, since the polymer structure is very susceptible to UV and ozone attack. The advantages and disadvantages of SBR sealants are given in Table 15.3.

15.6.   Nitrile Sealants and Small Joint Sealants

Nitrile rubbers can be used to make a solvent-based fluid caulking compound for use in small cracks. The rubber has good adhesion to various substrates and is generally formulated into two types of specialty sealants.

The clear sealant is used for very small cracks up to ⅛ inch wide, and the sealant is literally pumped into the crack using a modified oil can or a Plews gun. These sealants are recommended for use in sealing small openings in sash, masonry, and various construction joints of small width. This includes needle glazing around the perimeter of leaking channel glazed sash, sealing cracks in skylights, miter and butt joints, and cracks in brick, tile, and masonry joints as well as porcelain enamel construction. The sealant consists of essentially 62% polymer with the remainder being solvent. The openings that are sealed are either discontinuities in metal sash or shrinkage

TABLE 15.4. Advantages and Disadvantages of Nitrile and Acrylic Sealants for Small Joint Sealing

Advantages	Disadvantages
1. Very low viscosity	1. High shrinkage, up to 45%
2. Gunnable from oil can equipment	2. Very high cost
3. Good adhesion to metals and masonry	3. Application limited to cracks
	4. Only available in screw-top cans
4. Very flexible	5. Guns must be carefully cleaned with solvent
5. No primer needed	
6. Movement capability of ±12.5%	6. Pigmented colors: aluminum or black/bronze only
7. Long life	
8. Clear sealant used in cracks up to ⅛ inch	
9. Pigmented sealant used in cracks up to ³⁄₁₆ inch	
10. Good weather resistance	
11. Good UV resistance	
12. For needle glazing	

cracks that are not moving. The sealant adheres to all substrates, remains flexible, and is not affected by UV.

The pigmented sealant is usually made using a nitrile rubber, and has essentially the same solids content but is made using some filler. This sealant is recommended for slightly wider joints not exceeding ³⁄₁₆ inch. It is used in the same kind of applications, but the filler gives the sealant some structure and makes it non-sag in slightly wider joints. The sealant is not meant to take any movement, although the sealant dries to a rubber consistency and will take a small degree of movement. The movement capability would be of the same order of solvent-acrylic sealants, which is ±12.5%.

Small joint sealers can also be made using solvent acrylic either as a clear or a pigmented sealant. The properties would be the same. Nitrile-rubber-based small joint sealants might be more rubbery, but also higher priced. Any material selected for these restricted areas would be formulated so that it was easily gunable using a Plews gun into the narrow openings. The sealants are available in screw-top pint cans and larger sizes by special request.

The market is small, but the outlets are mostly in the construction market with some small over-the-counter segment. The materials are very expensive compared to conventional sealants, but nevertheless are a very necessary adjunct for any commercial sealant applicator of large scale, since complete job coverage requires a Plews gun and small joint sealant to complete the list of materials. Inevitably the applicator will find a small crack which can only be sealed with the thin viscosity small joint sealant. The advantages and disadvantages are given in Table 15.4.

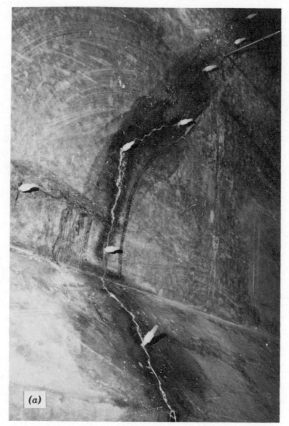

(a)

FIGURE 15.1. Epoxy repair. (a) One-way valves placed into holes drilled along crack line at 10-inch intervals;   *(continued)*

15.7. *Epoxy Sealants and Rigid Sealants*

Epoxy resins are used in construction for the sealing of cracks in structures and also to bond the two segments together. Figure 15.1 shows a structural concrete repair using epoxy resins. The repair concrete has greater strength than the original concrete if the repair is complete. If the repair will be subjected to water and water immersion, then epoxy must be used, since it is water-resistant. The epoxy compounds have also been used for adhering new concrete to old, and epoxy resins modified with polysulfide polymers introduce more flexibility into the system to take care of the difference in the expansion of the two substrates. Composite beams made by adhering new to old concrete have resulted in stronger structures.

Where the repair will not be subjected to water immersion or only dry application, then polyester resins will do a very satisfactory job. Both epoxy

(b)

FIGURE 15.1. (b) Quick setting epoxy gel seals crack surface and secures one-way valves; *(continued)*

and polyester have been used as concrete coatings for concrete bridge decks with satisfactory results, as covered in Chapter 28.

Epoxy flexibilized with polysulfide liquid polymers or coal tar can accommodate some joint movement up to a maximum of ±5%. Greater movements have been claimed, but these materials give a Shore A hardness greater than 80, and at this hardness they are not flexible.

Most epoxies are supplied as a two-component system that usually has very limited service life, ranging from minutes to probably 1 hour depending on mass and heat build-up. The reaction is very exothermic, and its own heat build-up greatly accelerates the cure. If the materials are mixed and immediately spread in a thin film, the cure rate is much longer. The epoxies are clear, and as such are used for electrical potting. For concrete work, the epoxies are filled with filler to lower the cost, and if the epoxy repair is to be large, then sand is used to extend the resin and also lower

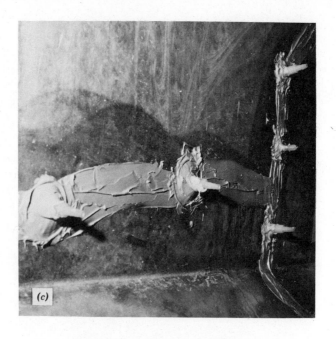

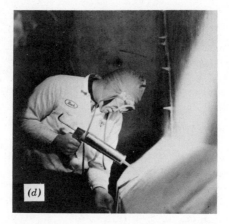

FIGURE 15.1. (c) Crack sealed, valves set, ready to grout; (d) Workman pumps high strength, low viscosity epoxy into lowest valve. After epoxy oozes from next highest valve, workman will move gun and repeat process until epoxy oozes from last valve. This assures complete penetration. (Courtesy of Sika Chemical Corp.)

the coefficient of expansion, since this repair could pop out in hot weather due to the much larger expansion of the epoxies.

Epoxy resins enjoy a large specialty adhesive market, but the volume as construction sealants is very small. The compounds are high-priced, with polyester being somewhat lower in cost. Both materials have to be mixed and used immediately, and crack filling and repair requires a special talent. Handling materials can be a problem since the amine catalysts for epoxy can be toxic and allergenic. Careful attention to mixing, handling, and application go a long way toward preventing problems.

15.8. Polybutene and Polyisobutylene Caulks

These polymers are variations of butyl rubber and have been discussed with the butyl sealants as possible additives to soften and plasticize butyl compounds and tape sealants. They also were discussed as tackifiers for other sealants.

The polymers have also been used by themselves to prepare noncuring gun-grade caulks and sealants that remain permanently tacky and semiliquid. The caulks can be made as a solvent-based caulk or as 100% solids material—which is desired in some areas where no solvent can be tolerated. The materials can be made with higher filler content and blended with some butyl rubber in various proportions to form uncured, semicured, and cured tapes, discussed in the next chapter.

The noncuring caulks are sold as curtain-wall caulks to be used in interlocking joints, for bedding of panels, sheets, rails, and moldings. They may be used between metal members as a buttering compound to keep out

TABLE 15.5. **Advantages and Disadvantages of Polybutene and Polyisobutylene caulks**

Advantages	Disadvantages
1. Good adhesion	1. Dirt pick-up
2. Remain plastic	2. Low cohesive strength
3. Self-healing	3. Non-elastic
4. 90 to 100% solids	4. Not recommended for moving joints
5. Excellent stability	
6. Good low-temperature flexibility	5. Poor solvent resistance
7. Excellent UV resistance	6. Can stain substrates
8. Excellent ozone resistance	7. No specifications
9. Good gunability	
10. Low cost	
11. Good water resistance	
12. One-component	

moisture. Some caulks remain permanently soft, and others may set to a Shore A hardness of approximately 10 after long-time aging. The caulks are made to remain gunable at temperatures down to 30°F. The caulks are best used in non-exposed areas since they remain tacky and will pick up dirt. They are very stable and will resist UV and ozone attack. Polyisobutylene is specially formulated at 100% solids with a small amount of carbon black for use as the primary seal in insulating glass because of its extremely low MVT (moisture vapor transmission). This caulk can be supplied as an extruded bead for hand application or in fillets for use in a high-temperature extruder for this application. The advantages and disadvantages are given in Table 15.5.

15.9. Roofing Caulks

These materials are generally bitumen-based and highly filled with fibrous materials. Some materials may be hot poured and perform both as adhesives and sealants. The hot-poured materials may be used as part of the built-up roof system, and the advantage is that upon cooling, the material is essentially set, since it is thermoplastic, and at 100% solids. Its use, however, requires an oil-jacketed kettle, so that the material is not recommended for small jobs.

The cold applied solvent-release material has greater flexibility and the material sets by the loss of solvent. A release of solvent from the surface results in a skin that inhibits the release of further solvent in the remainder of the sealant mass. Consequently, the body of the caulk under the skin may remain tacky for months. These cold-applied roofing materials are used for sealing and waterproofing cant strips, reglets, flashing, drains, and vent stacks during the roofing operation. Appearance is not a requirement and all materials are black. Cant strips might be embedded in the compound, and no shape factor is considered because movements are small. Tack-free time is on the order of 4 to 6 hours. The materials can become brittle if they do not contain high boiling hydrocarbons as plasticizers. The caulks are not elastomeric and have no recovery, but they will tolerate modest movement. They will exhibit some cold flow and will fuse together with time. The materials have good adhesion because they are semiliquid, and failure is usually cohesive, which may be the result of weathering and embrittlement.

Since the sealing of drains, stacks, cant strips, and so forth, is a part of the roofing operation, this sealing is included in the roofer's bond for the complete job. Consequently a service life of 15 to 20 years can be expected of a well-compounded roofing caulk. Other materials can do the job as well as the bitumen-based caulks, but in terms of price there appears to be no challenge. These materials are lower in price than most other sealants, and their modest share of the market appears to be relatively secure. Roofing caulks sell in the range of $2 to $5 per gallon, and the materials are usually sold in 5-gallon pails, or in chunks when used as a hot melt.

15.10. Sealants as Coatings

Although sealants are generally placed in joints, sealants can also be considered appropriate for some coating applications. Concrete structures are porous, and the surface of bridge decks is both porous and may contain fine hairline cracks that become focal points for further deterioration, such as freeze–thaw deterioration, which can widen hairline cracks and cause eventual breakdown of the concrete structure. Because water can lay on decks, it can seep into the concrete and cause corrosion of the reinforcing steel, which upon oxidizing expands and creates pressure for spreading the cracks. It is therefore imperative that the small hairline cracks and pores be sealed with some kind of sealant.

After several years of research, published reports by the Highway Research Board (18) show that linseed oil is a very effective seal coating for extending the life of portland cement concrete bridge decks and sidewalks. The linseed oil penetrates the pores and cracks of the concrete and seals them off, thus resisting deterioration by deicing solutions and the freeze–thaw cycle. There is some slight discoloration, and a tendency to dirt pickup as the linseed oil dries, but this does not affect the overall appearance of the structure. Spray coatings of silicone are also useful, but based on price, silicone cannot compete with linseed oil in bridge construction. Usually bridges should be resealed every year, since the oil film wears off or oxidizes and new material is needed to fill new cracks and form a new surface seal.

Epoxy and polysulfide-epoxy coatings have been evaluated and found successful, but at a prohibitive cost compared to linseed oil. The epoxy coatings were applied at a thicker film and abrasive spread into the film to form nonskid areas on bridges and approaches to toll booths. Such coatings were a two-part system that was mixed just before the nozzles, and complete cure was obtained in several hours to meet the tight time schedule for work on busy highways. The experimental work proved successful, but economics ruled out the products.

In building construction, where appearance may be more critical, the additional cost of a spray coating of silicone may be justified to preserve architectural beauty of a building. Elastomeric seal coatings have been applied directly to a concrete shell and dome roofs, but the success has been variable. The failures that have been reported are not due to poor coatings, but rather to an impossible stress situation. If a crack develops in the concrete, the coating bonded to the concrete must rupture. In actuality, the coating does have some extensibility, but a good rule of thumb is that the crack-bridging capability of any elastomeric film is equal to the film thickness of the coating. For this reason, buried membranes for membrane waterproofing are a minimum of 60 mils, which is ¹⁄₁₆ inches, and the film can bridge cracks up to ¹⁄₁₆ inch. In areas where performance is critical, a film thickness of 0.125 to 0.180 inches has been recommended to give wider crack bridging capability. On this basis, elastomeric coatings for domes should be as thick as the expected cracks, which might call for films up to ¹⁄₁₆ inch thick. In all probability, past failures occurred where only 5 to 10 mils of film were applied to give the dome adequate protection, yet no thought was probably given to the extent of cracking.

Where cracks have occurred in the surface of concrete and the concrete cannot be reamed out to make an adequate joint, then the use of a bandage design for joint sealant will keep out water and will take the movement, but at some loss in esthetics. This design can also be used on thin metal-to-metal joints that have developed leaks. Since the joint cannot be widened, this is the only technique to stop leaks. A description is given in Chapter 3.

Coatings are also applied to driveways and bituminous concrete roadways. These coatings are sometimes known as slurry seals, are fairly effective, and may be the only remedy for areas that show extreme "map cracking." These seal coatings may be asphalt emulsions or solvent-release bituminous materials. Slurry seals do a fairly good job of sealing cracks and enhancing the appearance of deteriorated asphaltic concrete. Their wear resistance is not outstanding, and in high-density traffic areas resurfacing is a better alternative. Bituminous slurries are quite inexpensive since they have a large amount of water and the bitumens are also inexpensive. They sell in 5-gallon pails varying in cost from $5 to $15 per pail depending on volume and quality. The materials set with the evaporation of water in a matter of hours. Although the market volume is high, there is no reference standard and quality is not controlled.

16

Preformed
Sealing Tapes

16.1. Introduction

Preformed sealing tapes were introduced to the construction industry in the early 1950s. The well-known "rope caulk" familiar to most homeowners began to appear in the local hardware stores about the same time. The tapes were slow in gaining initial acceptance, but their use has risen sharply. At the present time, the market compares to the market for urethane sealants and is growing at a faster rate. Part of the rise has been due to the increase in metal buildings that use tape in their assembly. The greater portion has been due to the large increase in glass in the construction market. The installation of glass requires glazing of various types depending on construction and design. Some metal designs are simple, but may require more complex caulking and glazing systems; other metal designs, although more complex, may only require two gaskets, one a compressible sponge gasket and the other a dense gasket.

The preformed tapes act as both sealant and resilient filler. Generally permanently tacky, they have become popular with contractors because they eliminate the need for complicated installation equipment and also eliminate waste. The job is also cleaner and quicker. In many cases the entire window is glazed from the inside, thus eliminating expensive scaffolding.

16.2. Compounding

Preformed tapes are produced in two general classes, resilient and non-resilient. However, with time several hybrids have appeared, such as a non-

175

resilient tape but with an insert of a cured butyl rubber rod. This is the preshimmed tape. Other versions might be classified as semicured, since they are both tacky and highly compressible but still exhibit some resiliency.

The nonresilient tapes are compounded from polybutene, or uncured butyl; combinations of butyl with polybutene; butyl and polyisobutylene; and blends of butyl, polyisobutylene, and polybutene to give nonresilient tapes of various textures.

The resilient tapes are compounded from either cured butyl or butyl modified with polyisobutylene, polybutene, and resins to give various degrees of hardness as well as surface tack. Fillers used are not high-quality since high physical properties are not required. The fillers might consist of clay, ground calcium carbonate, and some fibrous fillers for structure. Asbestos has been used in the past to help retain shape both before and after installation. Most tapes are generally black or aluminum gray, but other colors can be made. Although some tapes may contain solvent, most tapes are 100% solids.

16.3. Properties

Preformed tapes are essentially a strip of noncuring or partially cured sealant. Because they are noncuring and do not require solvent loss for curing, they have virtually no shrinkage after application.

The Shore hardness of preformed resilient tapes varies over a wide range, and may reach 30 to 40. Since the tapes are based on butyl rubber or related polymer for the most part, they do not exhibit much change in hardness when the temperature drops below 0°F. All the butyl-type tapes have excellent resistance to UV, ozones, and heat. Tapes conditioned to 175°F should not exhibit any oil exudation, blistering, flow, or loss of adhesion. Tapes used with plastic lights in windows should not contain any solvent.

Since the tapes do not cure, the normal physical properties for sealants do not apply. Nevertheless, the tapes are expected to have good adhesion to metals, glass, and masonry. Because they are used in glazing, the glazing tapes are expected to have excellent resistance to UV and ozone, and to maintain adhesion at room temperature and with water immersion.

Over the years the National Association of Architectural Metal Manufacturers (NAAMM) developed several reference standards. The SS-1c-68 standard was for nonresilient tapes, and SS-1b-68 was for resilient tapes. SS-1c-68 had requirements for hardness, oil migration, staining, low temperature flexibility, and adhesion. SS-1b-68 had requirements for hardness, adhesion, compression set, low temperature flexibility, corrosion resistance, shelf life, accelerated aging, and weathering.

Several years ago NAAMM stepped aside and left the specifications up to the Architectural Aluminum Manufacturers Association (AAMA). AAMA has adopted a number of reference standards that have served over the years to cover requirements for a number of glazing materials and sealants. Among these are three standards for tapes, namely:

AAMA 804.1, preformed tapes with limited structural strength. This is a replacement for SS-1c-68.

AAMA 805.1, preformed tapes with structural strength.

AAMA 807.1, preformed oil-extended tapes with limited structural strength. This is a replacement for SS-1c-68.

The above three standards all use the same test methods and differ only in the maximum Shore A hardness after heat aging. The three standards include requirements on weathering, hardness, adhesion, oil migration, sag, low temperature flexibility, ease of backing paper removal, and penetration. Some of the aging requirements are quite rigorous, calling for aging cycles of 21 days under an RS sunlamp, 21 days at 140°F, and 6 months of Florida sunlight exposure. Some adhesion tests require a 2-week water immersion before the test.

ASTM Subcommittee C-24.50 on Tapes has adopted a number of test methods in recent years and is presently working on a standard that will incorporate test limits for the various approved test methods. The test methods have a lot in common with both the NAAMM and AAMA standards, but also are more performance-related and will have more credence since they will be based on consensus. The test methods to be used in future standards for various preformed sealing tapes along with recommendations for requirements are as follows:

1. ASTM C-771, Test Method for Weight Loss after Heat Aging of Preformed Sealing Tapes. The test is run on a formed block of tape $1\frac{1}{2}$ × 5 × $\frac{1}{8}$ inches in thickness adhered to a sheet of aluminum and heat aged for one week at 212°F. The recommended weight loss should be 3.0% maximum.

2. ASTM C-765, Low Temperature Flexibility of Preformed Sealing Tapes. The test is run on a length of tape adhered to a thin aluminum plate and conditioned for 14 days at 158°F, and then placed in a low temperature unit for 4 hours at −10°F, and then quickly bent over a 1-inch mandrel in less than 2 seconds. There should be no cracking of the tape or any adhesion loss of the tape to the aluminum.

3. ASTM C-766, Adhesion after Impact of Preformed Sealing Tapes. This test is run in duplicate. A 3-inch length of tape is placed over a steel panel 6 inches long. With the release paper on the tape, the tape is covered with another steel panel and compressed with a 5-pound weight for 5 minutes. The cover plate and the weight are removed and one sample has the release paper removed from it. Both assemblies are then placed in a forced-draft oven for 14 days at 158°F and then conditioned for 1 hour at 75°F. The test specimen with the tape facing

down is placed over a 5-inch-diameter ring, and the 5-pound ball dropped from a height of 15 inches. If more than 50% adhesion loss of tape to steel occurs on either specimen, the specimen has failed.

4. ASTM C-772, Oil Migration or Plasticizer Bleed-Out of Preformed Sealing Tapes. Pieces of tape are plied together to form a pad $1\frac{1}{2} \times 1\frac{1}{2} \times \frac{1}{8}$ inches, leaving the release paper on the tape. The pad of tape is then placed on a sheet of filter paper and weighed down with a 2.2-pound weight for 5 seconds. Then two more pieces of filter paper are placed under the test specimen, and this assembly placed in a 158°F oven for 21 days. The filter papers are then examined for evidence of oil migration or plasticizer bleed-out. In this test various criteria must be set up, since for porous substrates a small amount of plasticizer may be needed for better wetting. Plasticizer is not needed if the tape is to be used on metal or glass.

5. ASTM C-782, Softness of Preformed Sealing Tapes. This is a test to measure softness of a preformed tape by using a penetrometer. The test is to be used for describing the desired softness by an applicator. Most tapes today do not give a penetrometer range, but this could change when a complete specification has been issued. Judging by available tapes today, some are stiffer, such as the unshimmed tapes for smaller lights. For larger lights a softer tape can be used since the tapes are pre-shimmed—which means that a small cured rod of butyl rubber is positioned inside the tape to control the desired spacing of glass to the stop. The small lights have a smaller size. Tapes are available with shims from $\frac{1}{16}$ to $\frac{1}{4}$ inch in diameter.

6. ASTM C-907, Tensile Adhesion Strength of Preformed Sealing Tapes by Disc Method. This is a test that measures the tensile adhesion strength of a preformed sealing tape. The test is run by first applying tape to one side of a round disc 1.597 inches in diameter, and then squeezing this against another disc until the tape thickness is exactly 0.125 inches. The tape and assembly is first conditioned to 77°F, and the finished assembly held for 15 minutes. Then the jig is pulled apart at a rate of separation of 2 inches per minute until failure occurs. The peak load and the amount of cohesive or adhesive failure are reported. Here again the value desired would vary quite widely, since there are soft resilient tapes and tough resilient tapes. Because the disc has an area of approximately 2 square inches, the values could be quite high, but could range anywhere from 10 to over 50 psi. Nevertheless, the test has considerable merit along with the other test requirements. With the softer tapes the failure could be cohesive, while with the stiffer tapes the failure would most likely be adhesive. It may be more logical to just separate the discs 100% above their expected maximum movement and require no adhesive or cohesive failure. It makes little sense to pull a soft tape to excessive elongation and then attempt to interpret the failure.

7. ASTM C-908, Yield Strength of Preformed Sealing Tapes. This test consists of squeezing 2 pieces of tape 3 × ⅜ × ⅜ inches to 0.2 inches between glass and a steel plate, holding for 1 minute, and then conditioning the specimen for 24 hours at 75°F. The assembly is then pulled apart at a rate of 1 inch per minute, and the yield strength reported along with adhesive or cohesive failure. Here again, since the contact area is over 2 square inches, the yield values could be high depending on the consistency of the tape; with soft tapes, the failures would be cohesive, while with stiffer tapes the failures might be adhesive.

8. ASTM C-879, Release Paper Used with Preformed Sealing Tapes. This test consists of taking samples of tape with release paper and rolling the tape several times with a 4.5-pound roller to ensure adequate contact of release paper with the tape. The samples are then heat aged for 14 days at 158°F. Other samples are conditioned for 14 days at 105°F and 100% relative humidity. The samples are removed and conditioned for 1 hour at 72°F. The release paper must strip off the tape after both exposures without any trace of compound to the release paper.

The test methods have considerable merit, and setting the requirements for ASTM C-907 and ASTM C-782 will be difficult, since the end use of the tapes varies considerably. For small windows, a stiffer tape is used, since the stiffness of the tape is needed to control the space between the glass and the stop. For larger lights, a pre-shimmed tape is used to prevent squeezing out of the tape, and the tape is softer to get better wetting. The cured bead of rubber, generally butyl, is quite tough, having a Shore A hardness of 40 to 60.

For other specific areas of application, the tape would have different handling and performance requirements. Curtain-wall tape may require better flexibility at −45°F. Metal building tape may require good handling consistency at 0°F. In many cases specific handling properties have been derived by consultation between the user and the manufacturer.

Finally, even though the various laboratory tests are very desirable and necessary, they may still not be sufficient to determine performance on the job site. For this reason, architectural firms have insisted on mock-up testing of typical wall assemblies including windows and glazing under dynamic water infiltration tests, buffeting wind and water, and static water infiltration. AAMA and ASTM have set up dynamic testing procedures for testing various wall assemblies. The tests may be run on any type of wall section, including granite, masonry, metal, and glass curtain wall with and without windows. The testing is done to suit the architect's desire to verify both weather and structural integrity.

The dynamic tests can evaluate the entire system including interior and exterior tapes, gaskets, and sealant, as well as the strength of the metal "stick" curtain wall if that is used. Tests run would include the following:

1. Dynamic water infiltration. Windows and wall subjected to water infiltration with slipstream velocity equivalent to 100 mph and water spray at 5 gallons per hour per square foot for 15 minutes.

2. Buffeting wind and water. Window and wall subjected to water spray of 5 gallons per hour per square foot for 6 cycles at 100 mph slipstream velocity wind for 1 minute, increasing for 30 seconds up to 137 mph, and then back to 100 mph. Test involves water on face of unit for 60 minutes.

3. Static water infiltration. Window and wall subjected to water spray of 5 gallons per hour per square foot and uniform static pressure up to 20 psf for 15 minutes equal to 90 mph. Total running time is 60 minutes.

4. Thermal and pressure loading. Window units heated and tested at 150°F to the following consecutive pressures to check the resistance to pump-out:

    a. Six cycles of minus 10 psf outwardly load for 30 seconds, then rapidly decreased to zero pressure.

    b. Repeat the above 6 cycles but with a minus 20 psf load.

    c. Six cycles of plus 10 psf increased load for 30 seconds, then rapidly decreased to zero pressure.

    d. Repeat the b cycle, above, but with plus 20 psf loading.

The water tests should show no leaking and the thermal and pressure loading should not show more squeeze-out than $\frac{1}{32}$ inch. Test facilities include the Construction Research Laboratory in Miami, the Industrial Testing Laboratory in St. Louis, and the Warnock Hersey International Ltd. in Toronto.

16.4. Application

It is in the area of job application that the preformed tape sealants really stand out. These sealants require no mixing, no special equipment, no messy clean-up, and are ready to use. The tapes are delivered to the job in neat rolls. They are packaged in cardboard boxes with centering cylinders and separators so that the tapes are not crushed or distorted. The tape is supplied with a nonsticking coated paper backing that permits rolling onto itself and packaging into rolls for easy storage as well as easy installation. The tapes vary in thickness from $\frac{1}{16}$ to $\frac{1}{4}$ inch, and the widths vary from $\frac{3}{8}$ to 1 inch.

Tapes are a more compliant form of preformed compression seal. They should not be used in working joints that have any significant degree of movement. Tapes are very satisfactory in joints having shear movements since the tape, being semifluid, deforms back and forth with ease. Tapes used in glazing are essentially subjected to shear movement, which does not present any problem if the lights are not too large. With large lights the tape can eventually be rolled out of the stop. For this reason, the stop may have a retaining nub to prevent the shim of the pre-shimmed tape from rolling out. Where very large lights are used, then a cap bead of sealant is recommended to keep the tape in place. Nonresilient tapes are used in the installation of residential storm windows and storm doors. The homeowner may use tapes for sealing cracks in masonry walls, bedding for base molding and shoe molding strips to prevent drafty floors, sealing around air-conditioning equipment, and caulking exterior doors and window frames.

The various tapes find their major application in glazing. There has been a greater effort made by the architect in designing windows with the desire

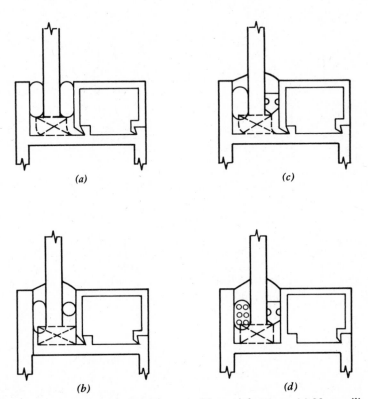

(a)   *(c)*

(b)   *(d)*

FIGURE 16.1. Wet-to-wet glazing systems with straight stops. (a) Nonresilient tape both sides for small windows up to 50 united inches. (b) Nonresilient tape with cap bead for exterior; a round roll-in neoprene gasket for compression with a cap bead for interior; good for any size window. (c) Same as (b) except with a wedge shim, good for any size window. (d) Closed-cell neoprene sponge gasket with a cap bead for exterior; wedge shim gasket with a cap bead for interior; good for any size window. (Courtesy of Tremco, Inc.)

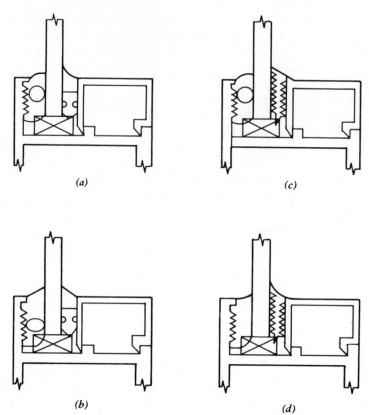

FIGURE 16.2. Wet-to-wet glazing systems. Exterior stop with nub; straight interior stop. (a) Pre-shim tape exterior; wedge shim gasket with cap bead interior; good for windows up to 150 united inches. (b) Pre-shim tape with cap bead for exterior; wedge shim with cap bead for interior; good for any size window. (c) Pre-shim tape exterior; vision strip in heel bead for interior; good for windows up to 150 united inches. (d) Nonresilient tape for exterior; vision strip in heel bead for interior; good for windows up to 75 united inches. (Courtesy of Tremco, Inc.)

to cut down on labor costs. Making it possible to do the entire glazing operation from the inside has also necessitated redesigning the metal hardware. This has also resulted in considerable saving in labor and time. The present glazing systems can be classified as one of the three following categories:

1. Wet-to-wet. This uses either sealant or tape on both sides of the glass. Sealant may also be used as a cap bead on either or both sides.
2. Wet-to-dry. This uses a sealant or tape on one side of the glass and a dense gasket on the other. The tape might have a cap bead.
3. Dry-to-dry. This uses gaskets on both sides of the glass. One gasket is a sponge gasket, which is compressible; the other gasket is a dense gas-

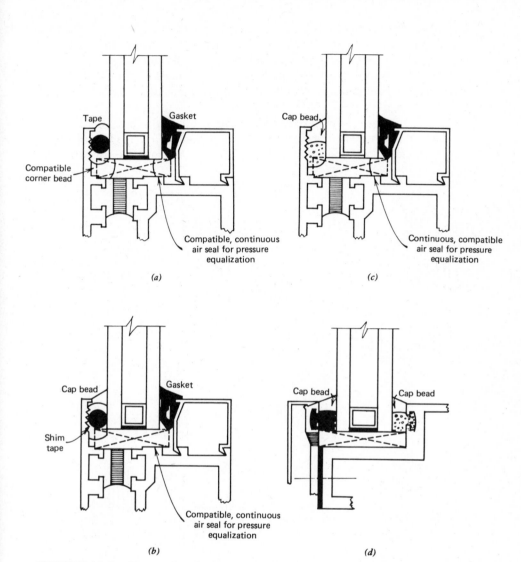

FIGURE 16.3. Wet-to-dry glazing systems; nubs or gasket races on both stops. (a) Pre-shim tape for exterior, lock-in wedge gasket for interior; good for any size window. (b) Pre-shim tape with cap bead for exterior; lock-in wedge gasket for interior; good for any size window. (c) Closed-cell neoprene sponge gasket with adhesive with cap bead for exterior; lock-in wedge gasket for interior; good for any size window. (d) Closed-cell neoprene sponge gasket with dart for exterior; dense compression gasket with dart for interior; both sides with cap bead; good for any size window. (Courtesy of Tremco, Inc.)

ket, which supplies the necessary pressure to make the system more water tight. Since there will be some water infiltration into the joint, this system is also used with weep holes. The architect has also designed the interior and exterior stops to receive gaskets with interlocking keys for easier installations as well as locking the gasket in place.

The choice of glazing detail depends on the window size and the details of the interior and exterior stops. There are any number of glazing combinations, and typical details will be given for the average situations. It must

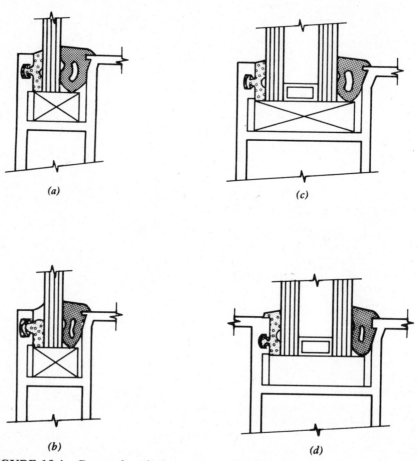

FIGURE 16.4. Dry-to-dry glazing systems. Gasket race in exterior stop: nub on interior stop; most popular systems. (a) Closed cell neoprene sponge gasket with dart for exterior; dense wedge neoprene compression gasket for interior; good for any size windows. (b) Same as (a) but with extra cap bead on exterior. (c) Same as (a) but for insulating glass. (d) Same as (c) but sponge gasket has a designed wide tab on exterior sponge gasket to deflect water. (Courtesy of Tremco, Inc.)

be born in mind that window and curtain wall manufacturers have hundreds of designs available depending on size, metal, metal design, and price.

The simplest metal designs with straight stops (Figure 16.1) require wet-to-wet glazing systems. Figures 16.1c and 16.1d would also work satisfactorily for Hope windows. Figure 16.2 covers glazing systems with a nub on the exterior stop and with a straight stop for the interior. These also require wet-to-wet glazing systems. The vision strips shown in Figures 16.2c and 16.2d are vinyl extrusions embedded in a heel bead that locks the strip in place. The vision strip is the forerunner of the lock wedge gasket. Figure 16.3 covers wet-to-dry glazing systems where the exterior and interior stops have nubs. Figure 16.3d is a hybrid system of dry-to-dry and wet-to-wet, since cap beads are used on both sides. Figure 16.4 shows typical dry-to-dry glazing systems requiring a gasket race to the exterior and a nub for a compression wedge gasket on the interior. Figure 16.4d illustrates a sponge gasket with a lip for better drainage to the exterior. Figure 16.4c and 16.4d are for insulating glass installations. Figure 16.5 illustrates two possible solutions for plastic glazing, where the plastic sheet is placed at the bottom of the opening rather than on setting blocks. Setting blocks are not needed since the plastic sheet is not prone to cracking with edge imperfections. Figure 16.5b uses dense gaskets on both sides.

In a case where a cap bead is used, silicone sealant is the preferred sealant today, although solvent-based acrylic sealant and polysulfide sealant have been used successfully in the past. For plastics, care should be used in selecting a non-solvent or a non-plasticized sealant since surface crazing can occur. Where insulating glass is used, care should be taken in determining

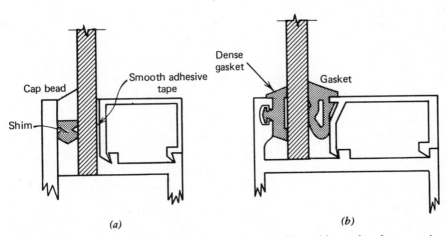

(a)   *(b)*

FIGURE 16.5. Plastic Glazing Systems. (a) Arrow shim with cap bead to exterior; smooth adhesive tape only on interior stop. (b) Dense neoprene gasket with dart to exterior; dense wedge neoprene gasket to fit nub for interior. (Courtesy of Tremco, Inc.)

possible incompatibility problems between toe and heal beads and between the various gaskets and the insulating glass sealant.

Although a number of glazing systems have been illustrated, other variations are possible. Where new situations arise, it is best to consult with a creditable sealant and gasket manufacturer who can run tests as well as have the systems tested under dynamic test conditions.

16.5.   Reference Standards

AAMA specifications 804.1, 806.1, and 807.1 have been discussed in Section 16.3. Some of the requirements are quite difficult—yet the specifications are not complete. The test methods derived by ASTM and also discussed in 16.3 are more meaningful and pertinent to the industry. While a specification is presently being derived, and will eventually be issued by ASTM Committee C-24, specification limits for several grades of glazing tapes are given in Appendix 8.

16.6.   Summary

Because of their easy installation, sealing tapes have experienced very rapid growth over the last 10 years, and seem to have reached a plateau in sales in the last several years. Growth has come mainly in glazing applications in the high-rise construction field. The large use of metal building tapes has also paralleled this fast-growing industry. Sealing tapes cost more than a sealant bead, but when labor is included the total cost is less for a tape. The total volume is probably as large as the urethane sealant market, with a good percentage going into the glazing areas due to the newer methods of installing windows. The design of tapes has kept pace with the requirements for various types of installation and window design. In other areas such as metal building, the market will depend on diversification in metal building design. The metal building design has led to faster and cheaper construction for all kinds of building including storage warehouses, gas stations, garages, and even small office buildings. This type of construction requires the use of a noncuring tape between overlapping metal sheeting. Another area is the use of various sizes and shapes of preformed tapes for a multitude of uses in miscellaneous light and heavy construction. The fact that most sealant manufacturers carry these tapes among their products attests to the popularity of this type of sealant.

17

Preformed
Gasket Seals

17.1. Introduction

Preformed gaskets are differentiated from preformed sealing tapes on the basis of cure. All preformed gaskets exhibit a good to excellent degree of cure and elasticity. When used, they depend on their elasticity and recovery to maintain a seal. Preformed gaskets can be dense rubbers as well as cured sponge gaskets of various compositions. The present gaskets come in a variety of shapes, hardness, densities, as well as compositions.

The concept of gasket sealing is as old as the problem of sealing. Any time man has had to mate two adjoining rigid parts in a leakproof fashion, a gasket is usually a first thought. In this broad sense, every seal can be considered a gasket. Cork, paper, rubber, and metal gaskets are used in automotive work, whereas in building work curable mastic type sealants are actually gaskets that are formed in place.

This chapter will be confined only to the preformed extrusions of high recovery that are used for construction sealing. These preformed seals are sometimes known as compression seals because they are placed into a joint under compression and rely on interface pressure to maintain a tight seal.

As early as 1931, gasket sealing was attempted in highway construction. The seal used was a simple rubber tube, similar to a length of a rubber hose. At that time, the seal was being placed along with a joint filler at the end of a slab unit. Considerable difficulty was encountered in keeping the seal in proper alignment, and therefore this attempt at gasket sealing was abandoned.

Preformed seals have been used in building construction for over 25 years. The prime application of preformed seals has been in glazing work.

One of the early developments in gasket sealing for buildings was an inflatable gasket made of synthetic rubber. This gasket fitted in a recess in the metal window frame. The sash was pivoted at the top and bottom to be opened for cleaning. When the gasket was inflated, the window has held tightly closed; when the gasket was deflated, the window could be swung for easy cleaning from inside the building. This type of gasket was successful, but expensive. Larger glass areas and less expensive extrusions forced it out of competition.

In the mid-1950s preformed compression seals were again introduced into highway and bridge construction. By the following decade compression seals had become the fastest-growing seal for highway and bridge construction. Currently, compression seals are probably used in more contracts for new construction than any other type of highway sealant. In the overall highway sealing market—comprised of new work plus resealing—compression seals rank second behind hot-poured sealants.

In building construction, preformed gasket seals have shown a steady growth, which is commensurate with the research and promotion money devoted to their development. Preformed seals certainly have many advantages, and a number of architects today specify them. Preformed seals are used mainly on large building jobs. They are seldom feasible for light construction or residential work.

Many types of rubber are now being used, but the predominant rubber is neoprene, with ethylene-propylene-diene-monomer (EPDM) challenging this position. Both materials can meet physical property requirements, but neoprene has fire resistance whereas EPDM will burn. With EPDM being cheaper, it is preferred if fire resistance is not a requirement. A third rubber with exceptional properties, but used only in limited areas because of price, is silicone. The silicone gasket is used where its exceptionally high recovery is required as in compression glazing, and also because the use of other rubbers might create a problem with adhesion due to migration of antioxidants to the silicone sealant used in stopless glazing. Vinyl chloride extrusions are used in the construction of insulating glass where a U-shaped channel may be used to cushion and hold the glass. The vinyl is limited to small size units mostly for the home and replacement market. Some vinyl extrusions have also been made and used in conjunction with heel beads in sealing smaller windows.

Butyl is used mainly as the cured shim in pre-shimmed tape because it is compatible with the polyisobutylene and polybutene compositions used in the preformed tapes. Other rubbers include hypalon and SBR, in low volume. Foamed vinyl chloride compositions are available in a wide variety of densities. Closed-cell gaskets with adhesive on one or both sides are available as lower cost gaskets for use in various areas of industry. Closed-cell urethane shapes with compressible features and relatively high densities are available as thermal stops for use in insulating glass. There materials have adhesive on both sides.

Another area where preformed gaskets seals are widely used is as a back-up for sealant in caulking joints. These materials are primarily closed-cell polyethylene foam rod and with very low density. A competitor in this area is open-cell urethane foam, which comes in low density and is usually rectangular and cut rather than extruded.

The various rubbers with their application areas, properties, and specifications are covered separately or in groups for specific application areas.

17.2. Properties

The properties of various synthetics used for preformed gaskets are covered in the following separate sections.

17.2.1. Neoprene and EPDM

Cured neoprene gasket seals were the first composition to be used when extrusions were designed. Neoprene rubber has been around a long time and has been shown to have excellent resistance to UV and ozone and retention of physical properties. Neoprene has been used as a wire coating and has established such an excellent reputation in performance that it will never be dislodged from many areas of application. The neoprenes can be extruded as both dense and cellular sponge extrusions. Where cellular extrusions are to be used as glazing gaskets, they are almost entirely made of neoprene. This may be primarily due to the easier and faster manufacturing of neoprene sponge gaskets.

In dense gaskets, there is fierce competition between neoprene and EPDM. Fire resistance is the qualifying feature of neoprene, and there may be an advantage where the gaskets are large and exposed. For small compression gaskets, the differences seem trivial; and EPDM is cheaper and can be formulated in colors whereas neoprene is strictly limited to black as a color.

The gaskets are delivered to the job site as fully cured materials, and there is no problem with odor or toxicity. The exterior gasket, which in a dry/dry gasket system could be the closed-cell neoprene sponge gasket, could come as a picture frame with molded corners. The corners have been the chief source of leaks, since butting and splicing of gaskets is always a problem. Premolded corners, even though more expensive, reduce the cost of installation, and the sellers of gaskets can design special molds that splice the gaskets by injection molding. The cost of the mold is added on to the cost of the gaskets, but for large jobs the increased cost per foot becomes negligible.

Neoprene closed-cell sponge extrusions are made with a thin impervious skin. The hardness is low and no value is given since it would be meaningless. The sponge extrusions are generally compressed 25 to 40% during installation, which will help to seal out the rain. The large amount of interest and activity in the glazing area has resulted in ASTM standard C-509 entitled "Cellular Elastomeric Preformed Gasket and Sealing Material."

The Shore A hardness of dense extrusions range from 40 to 75 depending on areas of application. Where the gasket is to be used as a continuous roll-in shim, the hardness is generally around 55 ±5. When the gasket is a wedge type, the hardness is generally 70 ±5.

ASTM C-864, "Dense Elastomeric Compression Seal Gaskets, Setting Blocks and Spacers," was derived to cover applications for glazing and sealing gaskets between mechanically restrained surfaces in building construction. This standard has two options. Option 1 calls for a flame propagation test, and Option 2 does not. Where no option is specified, then Option 1 will apply. This specification covers physical property and other requirements for 6 different hardnesses of rubber. The use of Option 2 permits EPDM to meet the specification. The setting blocks and spacers usually have hardness of 70 to 90 since the blocks must support the weight of the glass units and show little or no compression over long periods of time. Where neoprenes are used in highway and bridge joints, ASTM C-864 defines the physical properties for good-quality materials. The materials will be discussed in Chapters 19 and 20.

17.2.2. Butyl

Preformed butyl seals are thermosetting materials that can be extruded into a variety of shapes. The butyls can be made in sponge form, but because of higher compression set they cannot be subjected to as high movement as other rubbers. Butyls have excellent ozone and UV resistance but poor solvent resistance. Because butyl rubber is cheaper, this offers an incentive to use it, but its low modulus limits it from many areas of application. The largest application areas for butyl are inner tubes, tubeless tires, and tubing. Some butyl is used as the cured rod in pre-shimmed tape. Butyl as a gasket sees only limited use in building construction.

17.2.3. Silicone

Silicone gaskets are high-quality preformed seals. The silicones have excellent UV and ozone resistance and exceptionally high recovery. Their present use is restricted to gaskets in the use of silicone sealants for stopless glazing. Because gaskets are needed to support the glass, the gaskets must be neutral in their affect on the silicone adhesive bond. It was found that neoprene and EPDM gaskets contain antioxidants that can migrate and affect the adhesive bond. Silicone rubbers do not require any antioxidants; hence their desirability for this application. Silicone rubber is expensive and consequently is not used unless its exceptional properties are needed. This is another area where the properties are exceptional, but the price is also prohibitive.

17.2.4. Vinyl

Vinyl extrusions come in two forms, dense and foams. The foams are relatively new but have found wide usage as a gasket type for various installations. The vinyls are thermoplastic, but can be easily extruded to complicated shapes with good dimensional stability and control. The dense gaskets are formed over a wide range of hardness from 40 to 90. The lower hardness materials are plasticized and if the use application is low cost, the plasticizer will volatilize. Another problem with the plasticized vinyl is that the plasticizer can migrate to any sealant and effect its adhesive bond. This is a particular problem in insulating glass where, in the replacement market, a vinyl U-shaped gasket is used around the glass unit which is also sealed with a plasticized polysulfide sealant. It is imperative that the system be checked for compatibility. The plasticizers could migrate in either direction and cause adhesion failure or fogging of the insulating glass unit. Fogging is the result of plasticizer migration into the unit, with condensation on the inner cooler surfaces to form a haze. There are tests for fogging, as well as compatibility tests, and most sealant suppliers can make appropriate tests.

The vinyls are temperature-sensitive and become brittle at $-20°F$ and below. Vinyls have good resistance to UV and ozone, but are sensitive to plasticizer migration. Color retention is good even when subjected to weathering. Vinyls have good tensile properties and good tear resistance; however because of low recovery, materials should not be used in working joints. The seals exhibit a great deal of stress relaxation, and tend to relax and flow under sustained loads. They do not form an effective compression seal for working joints.

A recent development has been hot-melt vinyls for expansion joints on highways. The selection of proper high-temperature plasticizers has resulted in rubbery products for use in highway joints. The use of hot melts has simplified installation, which results in a rubbery seal as soon as the material has cooled—a matter of minutes. This subject will be covered in the chapter on highway joints.

One interesting product for glazing has been the design of a shaped vinyl extrusion with feet that can be anchored into a heel bead to give a finished glazing gasket for small lights, as shown in Figure 16.2.

Vinyls are very convenient to use. They can be shipped in 50- to 300-foot coils and require no special handling or storage. They can be easily cut to exact lengths and are subject to very little stretching during installation. Field splicing can be accomplished with a hot soldering iron. A V shape is cut out of the legs of a U-shaped gasket, and when the legs are folded, the corners close. Although the corners could be sealed with either a hot iron or a drop of sealant, this low-cost application probably receives no further treatment.

The use of vinyl foam seals has greatly expanded over the last 10 years. Today there is a wide range of seals with various densities and hardnesses (20 to 40 Shore 00 range for soft seals). The seals are all generally closed-cell structures with low water absorption, and are made to conform to various desired compression–deflection properties. The seals have low compression set if used within proper limits of compression. The seals come with adhesive on one or both sides, or with adhesive on one side and a polyester film on the other. Most seals are rectangular in shape, and can be supplied in various thicknesses and widths.

The service range is generally −20°F to 180°F. The volume for this type of seal is substantial and several large companies are involved in supplying these seals to industry. Areas of application include trailer sealing, computer cabinet dust and noise seal, vending machines, weatherstripping, auto and truck body frame seals, cab roof seals, corrugated panel seals, outdoor lighting fixtures, bus storage compartments, boat hatches, truck trailers, truck cabs, containers, computers, air-conditioning, refrigeration, door and window weatherstripping, garage doors, appliances, vibration dampening, and commercial refrigeration glazing. Although vinyl seals can take outdoor exposure, most of the seals are not exposed directly to the weather, but nevertheless a seal is required for various reasons. There are no ASTM standards for these products, but the products can be tested against a number of ASTM test methods for density, Shore 00 hardness, compression–deflection characteristics, compression set, water absorption, adhesion-in-peel, tensile adhesion, elongation, tensile strength, and weathering. In the applications mentioned above, the materials are used in high-priced equipment, and consequently the materials are also high quality.

17.2.5. Urethane

High density urethane foams are now available for specific outdoor applications. These are limited-use applications, but nevertheless the products meet the need. The densities are in the range of 30 to 35 pounds per cubic foot, which is high. Being urethane, the foams are highly resistant to wide temperature fluctuations. These seals have high tensile, high elongation, and high adhesive strength. Moreover, they also have low thermal conductivity. These seals are used as adhesives for wall displays, circuit boards, signs, auto component mounts, small appliances, tiles, weather stripping, kitchen utensils, mounting exterior trim, adhering letters and signs, skylights, as a thermal break for storm windows and doors, and insulating glass. All these seals have adhesive on both sides, and are tested against ASTM test methods including thermal conductivity.

17.2.6. SBR

Styrene-butadiene rubber (SBR) compression seals have not been widely used. The chief reason for the lack of acceptance has been their inferior weathering characteristics. Recovery is good and the cost is reasonable, but these seals have poor color retention and poor resistance to UV and ozone.

Consequently, they are not suited for use in exposed locations; they may, however, be used successfully where weathering is no problem.

17.3. Application

The application of preformed compression seals in building construction can be divided into four general categories:

1. Glazing seals.
2. Seals for exterior panels.
3. Structural gaskets.
4. Miscellaneous seals.

The following sections will discuss these categories in turn. The seals in each case are compression seals, which function by maintaining pressure against the joint panel. Preformed seals for highways and bridges will be covered in Chapters 19 and 20.

17.3.1. Glazing Seals

Glazing seals are manufactured in a multitude of shapes to fit specific applications (Figure 17.1). One of the most common shapes is the U-shaped channel gasket. These compression seals can be extruded from cellular neoprene, butyl, or (more recently) plasticized vinyl. Vinyl is the lowest-cost material and would be used in replacement windows and patio doors. The unit size must by necessity be limited since there is no provision for placing setting blocks as are needed for larger lights. Enjay designed a composite butyl gasket based on a cured butyl core with an outer layer of soft uncured butyl (Figure 17.2), which offered some interesting innovations, but this gasket received little acceptance.

The most widely used compression glazing gasket seals today are based on combinations of sponge and dense gaskets. These have been very popular because they permit the entire glazing system to be installed from the inside. While the materials can be more expensive, the labor costs have been lowered and the installation permits a complete one-shot application. The de-

FIGURE 17.1. A few examples of preformed seal shapes. (Courtesy of D. S. Brown Company.)

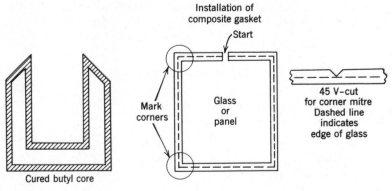

FIGURE 17.2.   Composite butyl gasket. (Courtesy of Exxon.)

sign of the metal stops permit a wide variety in the design of the glazing gaskets.

The illustrations shown in Figure 16.4 show three basic types of dry/dry glazing systems, each of which can be greatly expanded by the design of the metal stops and the gasket shapes. In each case, the sponge gasket is the compressible unit, which is installed at approximately 25% compression or is compressed by 25% upon insertion of the dense gasket. In some cases, an exterior cap seal of silicone sealant can be placed if desired to make a superior system. In all cases, the compression keeps most of the water out, but with wind buffeting and lower building interior atmospheric pressure some water may work its way into the building. The water tightness is greatly improved if the exterior gasket is made in one piece by using molded corners. The picture frame gasket is finished in the plant by molding all the corners in a specially designed mold for each particular shape of gasket. The gasket is usually made oversized by approximately 0.1 inches in each dimension, which places the gasket under a little compression for a tighter fit.

The better performing glazing gasket system is based on a wet/dry system using a pre-shimmed tape to the exterior and a compression gasket to the interior (Figure 16.3). These systems have been tested in finished window and wall assemblies at one of several test facilities throughout the country and have met dynamic and static testing proceedures in accordance with NAAMM requirements. These tests are described in Chapter 16.

The volume of glazing tapes and gaskets has grown tremendously over the past 10 years and will continue to grow as labor costs continue to mount. Very little replacement of gaskets in the dry/dry systems has taken place to date, but this system offers easy replacement if the gaskets take too much set.

17.3.2. *Seals for Exterior Panel Systems*

Compression seals for the one-stage weatherproofing of exterior wall panels are generally similar in cross-section to compression seals in paving joints (which are described in Chapter 19). These seals have not received wide acceptance because of installation problems. It is difficult to compress the seal and insert it into place properly in a vertical wall.

There has been some use of neoprene sponge gaskets of rectangular cross-section where the gasket has adhesive on one or both sides. The installation again requires careful attention. There has been use of a molded flat neoprene gasket similar to the waterstop design as an insert to make exterior waterstops in a rain-screen design. This gasket is inserted into a groove or slot in a masonry wall and its main purpose is to stop wind-driven rain. The equalized pressure within permits any penetrated water to flow down and out through the drains.

Most placement of gasket seals for exterior wall panels requires close tolerance to joint dimension, which is not usually the case, and this is where the system breaks down. Gaskets formed in place from a curable sealant solve most of the problems where misalignment can be a major factor.

A recent development resulted in a panel gasket designed to fit into a specially designed rigid PVC reglet with ribs into which a U-shaped neoprene gasket was fitted to give positive attachment by the use of integral splines (Figure 17.3). The panel gasket was designed to take 1¼ inch of movement while the opening was only 1 inch. This gasket was designed for use on a building in an earthquake area. The U-shaped gasket permits wide latitude in building tolerances and the interlocking feature can give considerable water tightness to the system.

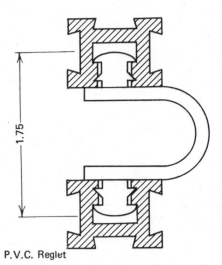

P.V.C. Reglet

FIGURE 17.3. Neoprene panel gasket with interlocking splines made to fit into a designed PVC reglet imbedded into the concrete to give a designed opening with controlled tolerances. The panel gasket can take up to 1¼ inch movement in a 1-inch-wide joint. (Courtesy of Standard Products Co.)

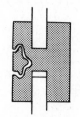

H Type perimeter

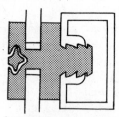

Intermediate supported
mullion/muntin

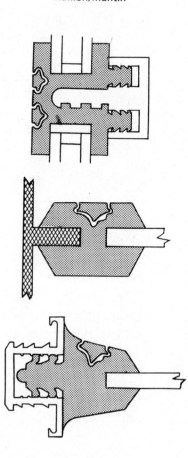

17.3.3. Structural Gaskets

The lock-strip gasket for architectural application is one of the dramatic new sealing materials that have appeared in recent years to meet the challenge of new structural methods. The concept was introduced in the late 1940s when the first major installation took place at the General Motors Technical Center, where more than 3600 glass and spandrel panels were glazed with lock-strip gaskets. With this introduction, considerable interest and a ready acceptance were given this novel glazing system.

The novel technique coined the term "zipper" gasket for these materials, which eventually were termed lock-strip gaskets. The industry went through many designs, some of which were very successful and others of which were not.

The technique permits the use of very simple metal or masonry design to which the gasket is attached, but in the end the structural gasket system is quite expensive since it is a bulky and heavy neoprene extrusion. The best technique is to mold the corners, and elaborate designs have been used to make multiple opening gasket systems. Examples of several types of structural gaskets are shown in Figure 17.4. One design shows a gasket placed in a rigid PVC reglet which has been set into the concrete and gives very accurate dimensions and interior ribs for better anchorage. Other modifications were made to permit better interlocking with supporting structural framing members. Dallen (19) gives an excellent treatment of the entire lock-strip glazing gasket technology. Special tools have been designed to make the lock strip easier to install. Figure 17.5 shows a new adapter handle which permits two-handed operation. Power-operated tools have been designed to assist insertion of the lock strip in cold weather. The new installation tool is driven by an electric impact hammer, which speeds installation. Figure 17.6 shows a complete installation using lock-strip gaskets.

The gaskets are meant to take a certain amount of movement, as shown in Figure 17.7. Design configurations resulted in certain basic requirements such as lip pressure, which should be a minimum of 4 pounds per linear inch. A hardness of 70 to 75 is needed in the gasket to supply the stiffness to hold the glass load. Other requirements are covered in ASTM C-542. This standard includes the test requirements and methods covered in Table 17.1

Structural gaskets went through a very active phase but in more recent years have waned in popularity. The very high cost for structural gaskets is one of the reasons, and leakage is the second. Over a span of time, the

---

FIGURE 17.4. Structural neoprene glaxing gaskets. Various designs of structural neoprene glazing gaskets. One design uses a rigid PVC reglet with ribs for better anchoring of the neoprene gasket. (Courtesy of Standard Products Co.)

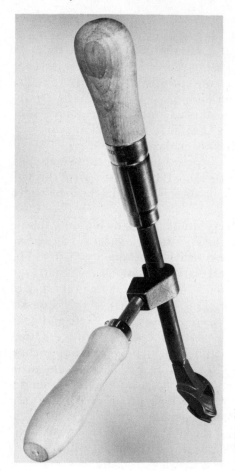

FIGURE 17.5. New adapter handle that permits two-handed operation in installing lock strip gasket. (Courtesy of Standard Products Co.)

lip seal pressure has fallen off, and the systems leaked during driving rain. The only solution was to apply a curing butyl sealant that would adhere to the plasticizer neoprene gasket and whose service life was only 3 to 5 years. Several major installations have been resealed several times although the appearance of the neoprene gasket is still very good.

17.3.4. Miscellaneous Preformed Seals

These seals come in a multitude of shapes for specialized applications. Some of the applications include pipe gaskets, continuous door stops, wiper-type gaskets for movable sash, gaskets for heating, and air conditioning ducts. Many of the special seal shapes are designed for plant application and may be part of the finished building component that is shipped to the job site. An example would be a large metal curtain-wall panel, complete with fixed-

FIGURE 17.6. Complete installation of lock strip structural gasket in an airport building. (Courtesy of Standard Products Co.)

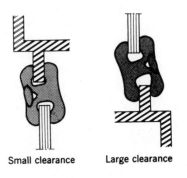

Small clearance   Large clearance

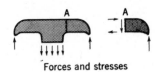

Forces and stresses

FIGURE 17.7. Action of structural gasket under load. (top) Action under wind loads. (bottom) Internal stresses.

199

TABLE 17.1.   **Requirements and Test Methods for Lock-Strip Gaskets**

Property	Requirements	ASTM Test Methods
Tensile strength, min.	2000 psi	D-412
Elongation at break, min. %	175	D-412
Tear resistance, min.	120 lb/inch	D-624 Die C
Hardness Shore A	75 ± 5	D-2240
Comp set, max. 22 hours @ 212°F	35%	D-395 Method B
Brittleness temp., min.	−40°F	D-746
Ozone resistance, ppm, 100 hours @ 100°F at 20% elong.	no crack	D-1149
Heat aging, 70 hours @ 212°F		D-573
Change in hardness, max.	+10 −0 duro points	
Loss in tensile strength, max. %	15	
Loss in elogation, max. %	40	
Flame propagation[a]	none	
Lip seal pressure	4 lbs/inch	C-542

[a]The requirement for flame propagation can be deleted if desired.

light glazing. One example of a field application of one of the specialized shapes is the cap seal shown in Figure 17.8. This cap seal forms a neat joint in exposed interior wall panels.

The market for cast-iron soil pipe is tremendous, and Smith (20), in his very informative chapter on cast-iron soil pipe, says that over 64 million gaskets were used in 1972. Approximately 50% of the volume used went for No-Hub gaskets for sizes from 1½ through 8 inches, while the remainder were compression gaskets from 2 to 15 inches in diameter. On today's market, the volume could easily represent over $20 million in sales. Figure 17.9a shows a mechanical joint that uses a gasket in compression and is

|←——1¼——→| **FIGURE 17.8.**   Cap seal. (Courtesy of D. S. Brown Co.)

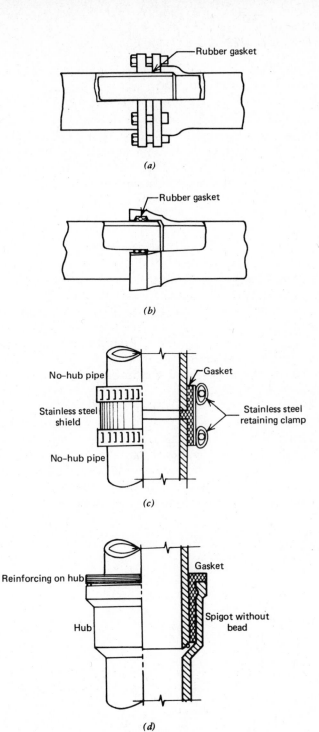

FIGURE 17.9. Various pipe gaskets. (a) Gasket for use in mechanical joints for cast-iron soil pipe. (b) Push-on joints for cast iron pipe. (c) Gaskets for no-hub joints. (d) Gaskets for compression joints.

used mainly for water pipe gaskets. Figure 17.9b shows a gasket for cast-iron pipe that is based on an O ring and used in push-on joints. The no-hub joint is illustrated in Figure 17.9c and Figure 17.9d illustrates a compression seal for soil pipe for use in hub and spigot assemblies.

This industry is supported by ASTM standard C-564, "Specification for Rubber Gaskets for Cast Iron Soil Pipe and Fittings." Although neoprene is the preferred material due to long time performance, other materials can be used if they meet the requirements of the test. Plastic pipe is making serious inroads into the home market for water and sewage pipe.

17.4. Summary

The preformed compression seals offer a generally high-quality product but also at a higher price. Preformed seals have a very good life expectancy, in the neighborhood of 20 years. These gaskets offer a good-looking joint with a minimum of job-site labor. There is little or no waste, a minimum of clean-up, and most jobs can be finished in a one-shot application.

The seals require close dimension tolerances, and openings must be ex-actly sized. Corners can be vulcanized in the shop to form a ready-to-use picture-frame gasket that also gives a minimum of leak problems after in-stallation for a small increase in cost.

The curtain-wall industry has grown rapidly and should continue to do so. The overall trend in construction is toward more prefabrication, which means more use of preformed gaskets. Other segments of industry also rely on preformed gaskets for a wide variety of uses. This market has grown considerably in the last 10 years and will continue to grow at the expense of some sealants. The trend is also to specialized gaskets and design.

18

Waterstops

18.1 Introduction

A waterstop is a specialized type of seal used for the waterproofing of concrete structures. The predominant waterstop is essentially a diaphragm cast as an integral part of the concrete to bridge the gap between concrete units. In recent years another waterstop has been introduced on the market in the form of a nonrigid rectangular extrusion. Waterstops are used in construction joints in walls and slabs—such as the corner joint where a concrete wall meets a foundation—and for many other uses. Waterstops are widely used in the industrial or heavy construction market. They are also used in heavy nonbuilding construction such as tunnels, dams, retaining walls, swimming pools, and culverts.

18.2. Materials

Watertops are a particularly demanding application. In building construction, if a sealant fails and the building leaks, the building can be recaulked. However, a waterstop is placed as an integral part of the concrete foundation construction and cannot be replaced. It must last for the entire life of the structure. Consequently, the choice of a material for a waterstop is a critical decision.

Waterstops have been fabricated from a wide variety of organic materials including PVC, neoprene, EPDM, SBR, bitumen, and other plastics and rubbers. Metallic waterstops are made from strips of copper, steel, stainless steel, and even Monel, and make good waterstops for many applications. The material used for a waterstop depends on the particular application. The waterstop is generally selected on the basis of the type and amount of movement it must accommodate. The chemical environment also determines the selection. Most waterstops are designed for low to moderate movement up to 15% of the joint width. However, particular designs are made to take more movement. Types of movement are about equally divided between tension and shear movements, depending on the particular design.

The vinyl (PVC) extrusions have become by far the most widely used form of waterstops. These PVC waterstops, like the preformed seals, can be extruded to very close tolerances. Typically, the PVC waterstops have good tensile strength (2000 psi), elongations of 300 to 400%, excellent tear strength (250 lbs/inch), and are in the Shore A hardness range of 70 to 85. The designed PVC waterstop is capable of some movement, particularly if the section between slabs is in the shape of a tube. No manufacturers have ever stated the amount of separation that can occur between adjacent slabs, but the expected maximum movement must be on the order of ⅛ to ½ inches, depending on bulb design or other features. PVC waterstops come in two grades, a standard grade flexible down to −20°F and a low-temperature grade flexible down to −50°F. Good low-temperature plasticizers with low volatility should be used in order that the waterstop remain flexible for the life of the structure. The waterstops embedded in concrete are generally unaffected by temperature, unless the structures are bridge decks and above-ground installations. PVC waterstops are most widely used where minimal movement is encountered.

Other elastomers such as neoprene, EPDM, and SBR can be used where movement can be encountered, such as isolation joints or control joints, and also the environment may determine the selection of rubber. Where there may be chemicals involved, such as in storage tanks, then neoprene may be desired. SBR may be used where storage or treatment tanks are involved, while EPDM might be considered where chemical resistance was not required. The rubbery materials require more elaborate extrusion equipment and they must also be subsequently cured, as compared to PVC, which is a thermoplastic. The extra handling increases the cost, which is probably the main reason for the popularity of PVC. Furthermore, the joints of PVC are field weldable whereas the rubber materials are not.

The metallic waterstops come in sheet form or coils. The steel, Monel, and stainless steel waterstops can also be obtained in different shapes such as ells, tees, and Xs in various sizes, as per requirement. Except for copper sheet, the other metals are not used apart from special cases where corrosion resistance is required. The copper can also be field welded.

A more recent waterstop introduced on the market in the early 1970s is a nonresilient material consisting of bitumen, which is supplied in tape form at approximately 1 inch square in 3 foot lengths and covered top and bottom with release paper. This waterstop is generally applied in a keyway, where the concrete is first primed and then the top release paper is removed just prior to the second pouring. The material is very tacky and permanently

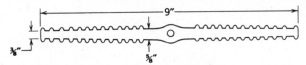

FIGURE 18.1. Waterstop with central bulb.

soft, adheres to both new and old concrete, and will withstand minimal movement. This material is laid end to end, and is considerably easier to apply than the PVC waterstops.

18.3. Application

PVC waterstops are fabricated in hundreds of individual shapes but with one central theme. The waterstops consist of two wing portions embedded in the concrete and a central portion to control the movement. The waterstop shown in Figure 18.1 has a hollow bulb in the center and is designed to accommodate moderate movement. The ribbed type with centerbulb is the most popular design with waterstops ranging from 3 inches to 12 inches in width. The centerbulb can vary from 3/16 to 1½ inches in diameter, and the web thickness can vary from ⅛ to ½ inches. These waterstops are claimed to withstand hydrostatic heads from 35 to over 200 feet. The serrations on the wings of the extrusion help bond the seal into the concrete.

Figure 18.2 shows a waterstop with no center bulb. This unit is designed for concrete joints that have little or no movement. The sizes, shapes, and thickness are very similar to the bulbed shapes. Variations on both types of waterstops include waterstops with a U-shaped centerbulb or a U center pleat. Some waterstops have reinforced ends for better anchorage, while other waterstops are a flat dumbbell with or without centerbulb. Some waterstops come with a split end, where the ends are split and nailed to the inside form. At the second pour the split ends are stapled together. This design was intended to prevent folding of the waterstop during the second pour. In addition, the Corps of Engineers has designed special V-shaped waterstops. New York State has designed L-shaped waterstops and troughs in addition to their own design for centerbulb waterstops. Other designs include a labyrinth and base-seal external waterstops. Some of the special shapes are shown in Figure 18.3.

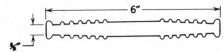

FIGURE 18.2. Waterstop with no central bulb.

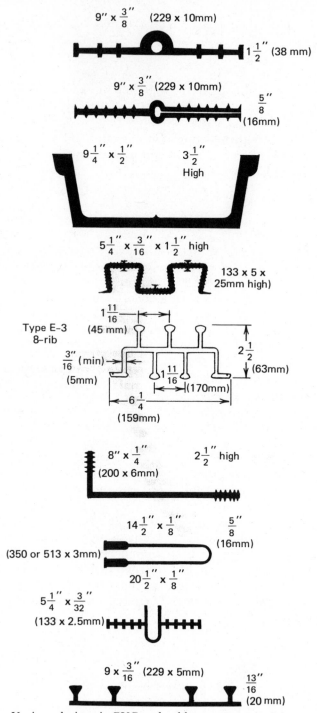

FIGURE 18.3. Various designs in PVC and rubber waterstops. (courtesy of Progress Unlimited, Inc.)

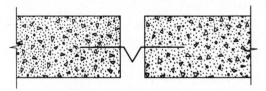

FIGURE 18.4. Copper waterstop.

Field join edges to form seal

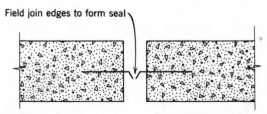

FIGURE 18.5. Field-joined copper or sheet vinyl waterstop.

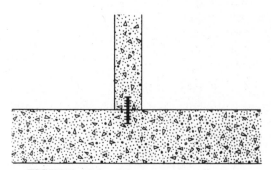

FIGURE 18.6. Waterstop installation.

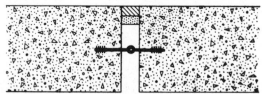

FIGURE 18.7. Waterstop in the base slab of a swimming pool.

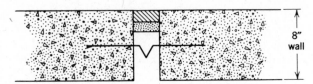

FIGURE 18.8. Section through the vertical joint in a retaining wall. Copper waterstop used with sealant.

Copper waterstops are formed from long strips of copper sheet (Figure 18.4). The wings of the waterstop are flat and rely on the bond between copper and concrete to obtain a waterproof seal. The movement between the adjacent concrete units is accommodated by the V-bend in the strip.

The waterstop may be completely responsible for the watertightness of a joint, or it may be used as a second line of defense in conjunction with another sealant.

The typical application of a waterstop involves embedding one wing portion of the seal in concrete and leaving the other wing portion exposed. As construction proceeds, this exposed portion of the waterstop is encased in the adjacent concrete placement, and the installation is complete. Waterstops of copper or vinyl sheet can be joined in the field, as shown in Figure 18.5.

A few joint details in which waterstops have successfully been used are shown in the following figures. Figure 18.6 shows a joint between a base slab and a foundation wall below grade. In this case, the waterstop is cast into the base slab with one wing portion exposed. As the foundation wall concrete is poured, it encases the upper portion of the waterstop. In this application, care must be taken to insure positive alignment and contour of the waterstop and to insure that it is not bent over and rendered ineffective by the weight of the fresh concrete placed on top of it. Note that in this application any relative movement between the adjacent portions of concrete places the waterstop in shear.

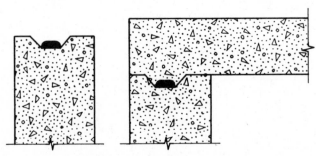

FIGURE 18.9. Preformed tacky bitumen waterstop for use in vertical or horizontal keyway joints.

Figure 18.7 shows the waterstop in the base slab of a swimming pool. In this application the waterstop is used together with another sealant. Relative movement between the adjacent slabs may either cause tension and compression due to thermal change, or shear caused by differential settlement.

Figure 18.8 shows a copper waterstop used with another sealant to waterproof the isolation joint in a retaining wall. Again, the movements may be either tension–compression or shear.

The nonresilient bitumen waterstop is generally laid in a keyway with the 3-foot sections laid end-to-end to form a continuous waterstop. The formed waterstop can be placed in a primed keyway after the release paper has been removed from the bottom and the top release paper removed just before the second pouring. The permanent tackiness of this waterstop maintains a seal against hydrostatic head. Figure 18.9 shows the waterstop in a vertical joint, and the same detail would work for a horizontal joint. Tests have been run showing that test assemblies have withstood hydrostatic heads. No data is available showing the amount of movement that the seal can take along with hydrostatic heads. Nevertheless, these waterstops have been used on many major jobs with apparent success and no leakage.

18.4. Summary

Waterstops can form an excellent barrier to prevent the movement of water through the many constructions joints in concrete structures. Once installed, waterstops cannot be replaced; consequently, material must be of the highest quality and installation procedures must be carefully supervised.

Waterstops are available in a wide variety of shapes. Materials most commonly used are PVC, with neoprene, EPDM, and SBR also used in specialty areas. Among the metallic waterstops copper sheet is most widely used. Other metallic sheets are steel, stainless steel, and Monel where special conditions are confronted. The sheets are used where more movement is expected.

A new waterstop made using bitumen has been employed with considerable success, and is easier to install in relatively nonmoving joints.

19

Sealants
in Highway Construction

19.1. Introduction

Sealing joints in highway pavements is one of the most troublesome problems facing highway engineers today. Although many joints are filled with tar, this is a very short-term solution that has to be repeated very frequently and is not acceptable on high-speed expressways. The annoying "thump-thump" of wheels over the pavement joints not only makes for a bumpy ride but renders vehicles harder to control at high speeds. The annoying thumps shorten the life of tires and suspension systems, and also costs millions of dollars in highway and bridge maintenance and repair.

Highway engineers are concerned with highway safety as well as keeping costs low for the life of the pavement. Properly sealed pavement joints will increase highway safety and also result in lower maintenance costs for the system. In spite of improvements made in sealing systems and technology, there is no easy answer, since different problems and climates exist all over the country, concrete compositions differ, substructures vary considerably, and available sealants are not always compatible with the spacing, movement, and design of the joint. Also, cost plays a major role in making the final decision for laying pavement and joint design.

Studies made on concrete roadways only confirm the complexity of the problem and one of the solutions is to recommend closer joint spacing. Both preformed gaskets and sealants have been used with success and failure, and a low-modulus silicone sealant has been evaluated successfully in a few trial areas. However, highway joints get considerably more abuse than any building joint, and the variables are more complex—which is the reason why highway joints will never be trouble free and the search for better materials and better construction will continue.

19.2.   History

"The proper sealing of construction, contraction, and expansion joints is a problem that has defied solution since the beginning of paved roads" (21). Early Roman roads were built of large aggregate and paving blocks that absorbed the movement over a large number of joints. In this country, the growth of the cement industry spurred the development of rigid pavements; as early as 1912, in fact, work was being done on joints. The first rigid pavements were built without joints and thus the pavements developed a random pattern of cracking. Consequently, the joint was introduced to control the cracking. In the early days, trial and error was the rule and experience the teacher. Early joint sealing materials included soft wooden boards, tar paper, sand, and tar.

The cement industry began to take an active interest in joint sealing in 1930 and the first efforts were made to provide a scientific basis for pavement joint sealing. Both expansion and contraction joints were used, dowels were commonly used, and definite answers were being sought to the problems of joint sealing.

By the late 1930s and early 1940s the state highway departments had acquired a considerable degree of sophistication. The states began to undertake studies of pavement movements and to record the performance history of pavements.

In 1960 the American Association of State Highway Officials began the most intensive program of highway research ever attempted. The results of this famous AASHO Road Test were published in several separate volumes. A great deal of work on pavement movement was included in this research program.

After World War II, the construction boom forced the development of more research into pavement behavior and also aided the development of newer joint-sealing materials. Rubber joint materials had been used as early as 1931, but had been largely unsuccessful in practice. Hot-poured asphalts were the most widely used sealing materials in the 1940s. The early 1950s saw the introduction of cold-poured elastomers. Polysulfides led the parade, to be quickly followed by other elastomers.

In recent years the old rubber tube concept of 1931 has been resurrected and given a new look. Preformed gaskets of neoprene have been used with considerable success in pavement joints. Another development is the "presealed joint." This device consists of two stainless steel plates and a neoprene gasket precompressed to a given width. This device is vibrated into the fresh concrete as the final step in the paving process.

In order to stimulate and fund highway research, the American Association of State Highway Officials, the Bureau of Public Roads, and the Highway Research Board have formed the National Cooperative Highway Research Program. This program has included several large-scale projects on highway and bridge behavior and also projects on joint sealing materials.

To coordinate the publication explosion, NCHRP sponsors the Highway Research Information Service, which provides an accurate record of research in progress and also furnishes a computerized search of the literature for research in any aspect of the highway research field.

19.3.   Why Seal the Joints?

Joints are expensive. They cost money to form and to saw, they are difficult and expensive to seal, and more expensive to keep sealed. One solution is to build roads without joints, such as the railroads with continuously welded rail a mile or more in length. In order to construct a pavement without joints or cracks, the pavement must be heavily reinforced. Some test sections have been built and some have been successful. The prime drawback is cost, since these pavements require a large quantity of steel and the highest quality of construction. The use of continuously reinforced pavements will continue but the cost will probably preclude its use in any but the highest quality of pavements. Prestressed concrete pavements have also been installed on an experimental basis in several states. However, cost will probably limit their use to the same applications as the continuously reinforced pavements. Consequently, jointed pavements will probably be with us for many years to come.

Joints in highway pavements are sealed to prevent the intrusion of incompressible solids, water, and highway deicing chemicals. All types of slab openings are sealed—contraction joints, expansion joints, longitudinal joints, and random cracks. Water and salt solutions seeping into the joint corrode the load-transfer devices between slabs, which inhibit normal pavement movement. Water seeping into pavement cracks corrodes the reinforcement,

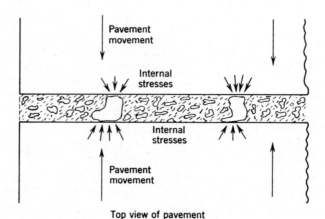

Top view of pavement

FIGURE 19.1.   Slab stresses due to incompressibles in the joint.

FIGURE 19.2. Pavement blowup.

FIGURE 19.3. Bridge seat tilted and cracked by pavement growth. (Courtesy of Acme Highway Products Co.)

which reduces the strength of the slab. Water passing through the joints also may result in the softening of the subgrade under the slab edges. This loss of subgrade support magnifies the pumping action of traffic passing over the joints and results in the cracking of slab edges and corners.

Solid material working its way into pavement joints is even more serious than water. In winter the slabs contract and the joints open. If not sealed, the joints fill with stones and road dirt. With the coming of warmer weather the slabs try to expand, but they have no place to go. Because the foreign matter filling the joint is not uniform in size, concentrated stresses are built up in the slab edges as the slab attempts to expand (Figure 19.1). These stress concentrations result in spalling of the slab edges, which accelerates the deterioration caused by water and deicing chemicals. Spalling is unsightly and makes the joint more difficult to seal in the future. Spalling is not confined to the tops of slabs. Many pavements are built on a granular sub-base; as the pavement expands and contracts, granular material is scooped into the joints and causes spalling. Studies have shown that spalling at the bottoms of pavement joints is at least as serious as the visible surface spalling.

Another problem associated with incompressible material in the joints

FIGURE 19.4. Joint filler extruded upward by pavement growth. (Courtesy of Acme Highway Products Co.)

is pavement growth. Pavement growth is undoubtedly the composite effect of a large number of contributing factors, but the joints are felt to be a large part of the problem. Incompressible material in pavement joints can prevent normal pavement movement. A pavement inhibited from normal movement may actually explode upward (Figure 19.2) or the whole pavement mass may move longitudinally. This pavement translation is not an isolated occurrence. A study by a research group at the University of Mississippi (22) showed that the problem exists at least to some extent in 80% of the continental states. Movements caused by this pavement growth may amount to as much as 1 or 2 feet. As an example, consider a pavement

FIGURE 19.5. Skew bridge pushed 2 inches out of alignment by pavement growth. (Courtesy of Acme Highway Products Co.)

with joints spaced at 60 feet. This spacing gives 88 joints per mile of pavement. If just ⅛ inch of incompressible material finds its way into these 88 joints, the total movement that must be accommodated is 11 inches.

Pavement translation may result in the displacement and misalignment of curves, or it may be accommodated at the nearest bridge structure. The moving slab often tilts and cracks the backwall of bridges, pushes bridges out of skew, or in extreme cases actually pushes the bridge off the bridge seat. Figures 19.3, 19.4, and 19.5 show some of the damage caused by pavement growth.

19.4. Types of Joints

Joints in pavement construction fall into two general categories, contraction joints and expansion joints. The contraction joint is a weakened plane in the slab, which is produced either by forming or sawing. The depth of the cut is usually ⅙ of the slab depth. The expansion joint is a full depth joint through the slab. Construction joints, which occur at the end of a day's paving operations, are full-depth joints through the slab. A day's construction is generally terminated at the end of a slab, so that the construction joint is, in effect, another expansion joint or contraction joint depending on state practice.

19.4.1. Contraction Joints

The contraction joint (Figure 19.6) is a transverse joint sawed or formed to create a weakened plane in the slab. The opening is generally ⅙ of the slab depth. As the drying shrinkage takes place in the concrete, the slab cracks throughout the remainder of its depth. After the cracking has oc-

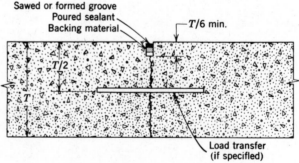

FIGURE 19.6. Contraction joint.

curred there is still some aggregate interlock, but usually a load-transfer device is placed in the slab to prevent relative vertical movement between slab edges. Several patented types of load-transfer devices are available (Figure 19.7), but the straight dowel bar (Figure 19.8) is still the most common type of load-transfer device in use. The contraction joint is generally ¼ to ½ inch wide; the ⅜-inch joint is the most common. The width of the joint should depend on the joint spacing; the longer the slab unit, the wider the joint opening should be. Joint spacings also vary quite widely from state to state. New York State uses a spacing of 60 feet 10 inches, and California uses a staggered spacing of 13, 19, 18, and 12 feet. The staggered spacing is effective in reducing any rhythmic thump under the wheels of vehicles. Several states are also experimenting with skewed contraction joints to distribute wear and reduce wheel thump.

Pavements nowadays are seldom constructed one lane at a time. Large slip-form pavers can pave sections over 24 feet wide. Consequently, a longitudinal joint is sawed in the pavement to separate lanes. This longitudinal joint is effectively another contraction joint. Its depth may be the same as the transverse contraction joint, and dowels are spaced along the joint to transfer load.

The contraction joint is intended to accommodate the movement of slab spacing. The joint is sealed at the top only. If a bulk sealant is being used, a back-up material is placed under the sealant to maintain the proper shape factor (see Figure 19.6). If a preformed seal is being used no back-up material is needed, but the saw cut may be stepped to seat the seal properly (Figure 19.9).

FIGURE 19.7. A patented load-transfer device. (Courtesy of Acme Highway Products Co.)

FIGURE 19.8. Dowel bars in place at a pavement joint. (Courtesy of Acme Highway Products Co.)

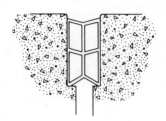

FIGURE 19.9. Stepped saw cut with preformed seal.

218

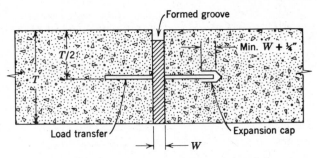

FIGURE 19.10.   Pavement expansion joint.

19.4.2. Expansion Joints

Expansion joints in pavement function in a slightly different fashion from isolation joints in buildings. The isolation joint in a building separates the structure into two units. Ties or keyways may connect the parts, but the requirements differ from the pavement slab. The expansion joint in pavement must permit the movement of the pavement in a longitudinal direction, but it must prevent the relative movement of the slabs in either the lateral or the vertical directions. In addition to the forces caused by expansion and contraction, the pavement joint is subject to the constant pounding of automobiles and heavy trucks. Considering a traffic density of 10,000 vehicles per day—which is not unusual—the joint takes quite a beating.

The expansion joint is a formed joint. It is a full-depth joint through the slab and is wider than the contraction joint. Expansion-joint width should depend on expansion-joint spacing. Practice varies widely from state to state. Many states use only contraction joints and no expansion joints except at bridge approaches. Other states use varied spacings for the expansion joints. A relatively common practice is to space expansion joints at approximately 1000 feet. Width of the expansion joint varies from ¾ to 2 inches. Since the expansion joint is a full-depth joint, a joint filler is usually placed into the joint from the bottom of the slab to within 1 inch of the top. The expansion joint is then sealed in the same fashion as the contraction joint. The joint fillers vary: redwood boards, cork boards, and asphalt-impregnated fiberboards are all widely used. The expansion joint, because of its greater width, is generally heavily doweled (Figure 19.10).

19.5. Joint Forming

The joints in rigid pavements may be made by inserts in the pavement, by sawing, or by hand forming. Hand-formed joints were the first type to be

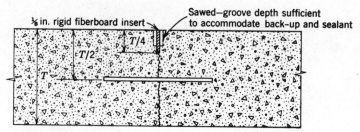

FIGURE 19.11. Contraction joint with sawed-out insert.

used in pavement construction. However, with the advent of the paving-train concept of pavement construction, it is not uncommon for paving crews to place more than a mile a day of two-lane pavement. At this rate of construction, hand finishers simply cannot keep up with the rest of the paving operation and, consequently, sawed joints have become the most widely used method of joint construction.

19.5.1. Hand Forming

Before the sawing of joints became practical, hundreds of miles of very good pavement were placed with hand-formed joints. Hand forming is still widely used in smaller paving contracts. In the hand-forming method, the cement finisher simply uses a hand-edging tool to create the weakened plane in the slab. The method does have the inherent disadvantage that the concrete at the joints may be overworked. This overworking results in

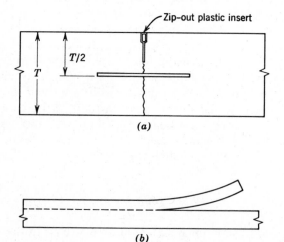

FIGURE 19.12. (a) Contraction joint with zip-out insert. (b) Zip-out insert. Top portion is removed to form groove for sealant. Lower portion remains in pavement.

weak joint faces that contribute to spalling and adhesive sealant failure. However, skilled finishers can do an excellent job of joint forming without overworking the concrete.

19.5.2. Inserts

Inserts have been used to form joints for over 30 years. Dozens of types have been tried, but none have been very widely accepted for any period of time. The most common types of inserts in current use are the "sawed-out" insert and the "zip-out" insert. The sawed-out insert is a thin sheet of hard fiberboard. It is secured into position before the concrete is placed (Figure 19.11). The advantage of the sawed insert is that the concrete is allowed to crack below the insert and thus the time of sawing is not so critical. The zip-out insert, generally made of plastic, is fabricated with a tear line. The concrete is allowed to crack and the top portion of the insert is zipped off, leaving the joint ready for sealing (Figure 19.12).

19.5.3. Sawed Joints

The time and manner of pavement-joint sawing is extremely important to both joint performance and sealant performance. The time of sawing is related to the curing of the concrete. If sawing is done too soon, the concrete will not have attained sufficient strength. This will result in aggregate pull-out and ragged joints that are difficult to seal. If sawing is delayed too long, the shrinkage cracks will have already formed in the pavement. Another danger with late sawing is that cracks will appear ahead of the saw blade, resulting in ragged joints. For normal portland cement concrete, the sawing should be done 48 to 72 hours after placement at 75°F. In hotter weather, the sawing may be done earlier.

Saw cuts in pavement may be formed in several ways. One convenient way to saw contraction joints is to use a single blade in one pass and make the full-depth (T/6) cut. A second pass can then be made with a "gang blade," or two blades and a spacer. This second cut saws the joint to its proper width and need only be deep enough to receive the joint sealant.

19.6. How the Slabs Move

Pavement slabs are generally assumed to expand and contract longitudinally as a function of temperature. Actually, the pavement movement is quite complex. Pavements expand in summer and contract in winter; hence, the joints open in winter and close in summer. Superimposed on this yearly cycle is a daily cycle of movements. The joints open in the cool hours of the night and close under the warm daylight sun. Research studies of the daily pavement cycle show that the movement of the pavement generally parallels the midslab temperature and not the slab surface temperature or the air temperature. The curves for temperature and movement, when plotted together, show that the peaks and valleys in the movement curve for the slab are roughly 2 hours behind the corresponding peaks and valleys in the temperature curve.

Pavements do not move smoothly. As temperature goes up, the slab attempts to expand, but is restrained by the friction between slab and subgrade. When the expansion force has built up so that it exceeds the friction, the pavement moves. This movement is fairly rapid, but usually not a sudden lurch of the pavement. Friction in the load-transfer device accentuates this stick/slip type of movement. If the load-transfer system in one joint "freezes" because of misalignment or corrosion, the adjacent joint must accommodate twice the normal amount of movement. This joint freezing will result in sealant failure, regardless of the type of seal being used.

Pavement curl is another factor that complicates the overall movement pattern. Curl is caused by the differential in temperatures between the top and bottom of the slab. Under a hot sun, with the air temperature at 85 to 90°F, the surface temperature may be as high as 125 to 130°F; whereas the bottom of the slab at the same time may be 75°F to 80°F. This temperature differential between the top and bottom of the slab causes the top to expand more than the bottom so that the slab arches or curls upward. In the cool night hours the process is reversed. The top of the slab is cooler than the bottom and thus contracts more, causing the slab to dish upward at the edges. This uneven movement of the slabs resulting from pavement curl may cause the load-transfer devices to bind, especially if long slab units are used.

The dishing upward of slab edges also causes the slab to have less subgrade support at the joints, which contributes to the cracking of the slab ends and corners under the pounding of heavy truck traffic. Such movement has even caused the slabs to break in half. Recognition of the problem of pavement curl has caused many states to use shorter joint spacing to minimize the problem.

Studies on test pavement illustrate the vast differences that can occur between the real and the theoretical. Minkarah (23) ran a very complete study on a section of highway in Ohio. The test section was approximately 0.8 miles long of reinforced portland cement concrete with joint spacings of 17, 21, and 40 feet. A number of variables were introduced, including plain and plastic-coated dowels. Minkarah found a correlation between midslab temperature and movement. Temperature variation across the depth of the slab was nonlinear. No correlation could be found between joint movement and air temperature, slab surface temperature, or subgrade temperature. This study was made using preformed seals that performed satisfactorily. In the particular joint design, the author claimed that sealant would have failed.

Moisture absorption also affects the size of concrete slabs. The difference in size between a dry and a wet pavement may be as mush as the total yearly thermal expansion. However, the pavement on grade and the moisture content varies from the top to the bottom of the slab. The bottom of the slab is seldom completely dry. At the present time, there is no accurate method of determining the actual movement due to moisture, and this factor is usually neglected.

Under the action of truck traffic, there is a relative vertical movement of the slab edges. In the warm parts of the day, when the center of the slab unit is arched upward and the slab ends are forced down into the subgrade, the relative movement of slab ends is difficult to measure and may be one or two thousandths of an inch at the most in a relatively new slab. In the very early hours of the morning, however, when the slabs ends are dished upward, the movements are 10 times as great as in the warmer parts of the day. The total movements in a new slab pavement are still quite small, but there is enough vertical movement under truck traffic to cause fatigue in a sealant. As the slab ages, the vertical working of the slab ends causes the dowels to "pocket," so that the vertical movement of slab ends tends to increase.

The movements of slabs may be calculated by using the temperature range, slab length, and coefficient of expansion for concrete. However, a quick rule of thumb is that the highway slabs on grade will move $1/16$ inches for each 10 feet of slab length throughout the normal temperature range prevailing in most of the continental states. Although generalizations can and have been made on slab movement, probably the biggest problem is that extremes in temperature do occur. All sections of the country periodically have very high temperatures or very low temperatures for extended periods of time during which the pavement gets unusual exposure to the weather. It is probably at these times that some of the initial failure occurs in an expansion joint, which is then aggravated with further movement. Add to this the variables of aggregate, subgrade differences, and joint and slab design, and the problem become real. It is impossible for the highway engineer to design around these extremes, and consequently compromise is the result.

19.7. Materials Used

Many different materials have been used in an attempt to successfully seal highway pavement joints. The material in the joints must be hard enough to prevent the intrusion of foreign matter into the joint (Figure 19.13), and should also prevent the passage of water through the joint. In addition, the sealant must have excellent resistance to the deicing salts used on pavements.

The history of paving sealants has been a stormy one. The material used

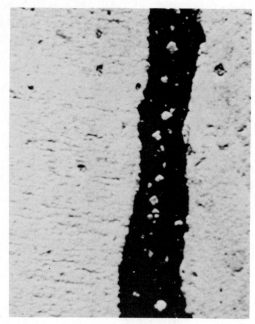

FIGURE 19.13.   A close-up of a pavement joint, showing stones and dirt embedded into the sealant.

in more joints than any other is hot-poured asphalt sealant. As recently as 1975 70% of the states still permitted the use of this material. This type of sealant, if well compounded and properly handled in the field, can do a good job of sealing pavement joints.

In the 1950s polysulfide sealants were introduced to the highway market, but did not meet expectations because of poor aging performance. The polysulfides hardened and failed in adhesion. Urethanes were also tried, with not much better results. Both polysulfides and urethanes were plasticized with large amounts of coal-tar derivatives, and the problems were getting good flow while maintaining low volatile content, as well as keeping the cost reasonable. The resulting failures caused many highway engineers to turn a deaf ear to the use of elastomeric sealants for highways. In 1966 only 10% of the state highway departments permitted the use of elastomeric sealant. In recent years, two materials have emerged as elastomeric sealants for highways: hot-melt PVCs and low-modulus silicone sealants. Hot melt PVCs have evidently proven themselves in the 1970s and early 1980s when used with carefully designed concrete pavement. Low-modulus silicone sealants have been tested in highway pavement joints in several successful installations.

Preformed neoprene gaskets have been used in both highway and bridge deck joints with good success. The compressions seals function by exerting a pressure against the faces of the joint at all times. In 1966 only 10% of the state highway departments permitted the use of compression seals. By 1969 the figure had doubled, and further growth has taken place since.

Neoprene seals do the best job of keeping incompressible solids out of the pavement joints. They also provide the neatest-looking joint of any known sealant. Compression seals keep a large percentage of water out of the joints, but are not truly watertight.

Compression seals, however, have several disadvantages. They cost 5 to 10 times as much as poured seals. This cost differential may be amortized if the compression seals can function effectively for 10 to 15 years without resealing. Compression seals also require straight, firm joint walls in order to function properly. Furthermore, spalls are difficult to repair. The spalled pavement edge must be rebuilt to its former line and grade, and a new compression seal must be installed.

The presealed joint, consisting of two stainless steel plates and a pre-formed compression seal, is still in the development stage and has not been used to any significant degree. Figure 19.14 shows a test section being installed.

Cold-poured bituminous sealants are rapidly diasappearing from the highway scene. Very few highway officials express any interest in continued use of this material.

FIGURE 19.14. Installation of a presealed joint.

Busch (24) describes the use of elastomeric concrete in expansion joint transition dams in bridges. No mention is made of the actual elastomer used, but the material uses an elastomer as a binder along with 75 to 80% of dispersed aggregate. Evidently this material is not meant for the average expansion joint.

19.7.1.   Hot-Poured Rubber Asphalts

The hot-poured asphaltic materials have undoubtedly been the most used and most misused of any highway joint sealant. The original materials were straight asphalt. In an effort to upgrade the material, finely ground rubber was added. The rubber was obtained from cleaned ground-up devulcanized used tires. The early asphalt sealants contained approximately 25% ground rubber and were good-quality sealants. As competition increased, the percentage of rubber began to drop, so that by 1967 few rubber asphalt sealants contained as much as 5% rubber. A new and somewhat tighter specification has forced the upgrading of these sealants since 1967.

The hot-poured rubber asphalt sealant should be made from good quality asphalt and should contain at least 25% ground rubber. The rubber should be SBR and should be uniformly ground to a 30 mesh size. Hot-poured rubber asphalts, using neoprene and other synthetic rubbers, are currently being manufactured. A great deal of both laboratory and field testing remains to be done, but the neoprene asphalts look very promising.

The hot-poured sealant must be carefully handled at the job site. The sealant is furnished to the job site in solid form in large drums. The material must be heated uniformly to a melt, so that it can be extruded into the joint. The heating should be done in a jacketed oil-bath heating kettle. Temperature should be carefully maintained. A temperature of 400°F can safely be tolerated, whereas temperatures over 450°F will "burn-out" the material so that it will not function properly. Also, these materials should not be reheated from one day's work to the next.

The hot-poured sealants, when properly placed, can effectively seal joints for a period of 5 years. These sealants are rubbery materials and should be placed into clean, sound joints with a reasonable shape factor. The optimum 1:1 width to depth ratio in the joint should be used whenever possible. However, these are low-recovery sealants, and the depth of the sealant may be greater than the width, if necessary, in order to obtain sufficient adhesive area. In the normal ⅜-inch-wide contraction joint, the depth of the sealant should be ½ inch. This shape factor can be obtained by the use of butyl rod stock and a bond breaker placed into the joint before installation of the sealant.

The hot pours have good adhesion to concrete without the use of primers. These sealants have only fair recovery properties, but good resistance to deicing chemicals. They are especially useful for resealing work and for crack sealing in both rigid and flexible pavements. The hot pours are also useful for sealing badly spalled joints in older pavements. In this application both the preformed seals and the elastomers are useless.

A hot-poured sealant can be installed in a pavement joint for approximately 30 to 40 cents per foot of joint. At this price, even if resealing is required every 5 years, the sealant is still a good buy. This material also has the advantage of familiarity. Construction and maintenance crews know the material and the equipment necessary to install it. Many state highway crews and contractors now have modern truck- or trailer-mounted equipment capable of sealing 20 feet of contraction joint per minute.

19.7.2. Cold-Poured Elastomers

The only elastomeric sealants used in highway joints are polysulfides and urethanes; silicone sealants are being experimentally evaluated. The polysulfides were the first to be used and rocketed to a large volume for a period of 5 years, and then fell off sharply as they began to show bad aging characteristics. The material looked good in the laboratory, but price dictated the formulations in the field. The use of relatively volatile coal tar hydrocarbons to give easily handled materials resulted in the poor aging after volatilization of some of the coal tar. Evidently the market would not accept conventional sealants at higher price and higher polymer content and the market greatly decreased within a few years. Urethane technology was similar, but never enjoyed the popularity that polysulfides did in their prime. Since 1966, some manufacturers have produced and marketed higher-quality polysulfides and urethanes that perform better. Because of the early failures, some of the market swung over to preformed compression seals.

The elastomeric sealant for use in highway joints should be a two-component material that can be accelerated to cure in 1 to 2 hours to give a cured joint sealant. The two-component materials should have a Shore A hardness of 15 to 25 in order to resist penetration of stone and road dirt. The one-component materials are not practical. Joint design should conform to a joint sealant movement of ±25% from the time of installation. When the slab length is 40 feet, the minimum width of joint should be ¾ inches, which will accommodate approximately ±³⁄₁₆ inches of movement. If the movement is greater, then the joint should either be widened or else compression seals should be used—which will accommodate greater movement capability on the order of ±35% and more.

The two-component elastomers for highway use may be either hand-mixed or machine-mixed materials. The hand-mixed materials are furnished in the usual ratio of 15 parts of curing agent to 100 parts of polymer base. These materials are mixed by an agitator blade using a slow-speed electric drill, and may be either poured or gun-extruded into the joint. The machine mix materials are furnished in a 1:1 volume ratio. The two components are placed in separate chambers on the mixing machine, pumped

FIGURE 19.15. Trailer-mounted sealing machine for two-component elastomers. (Courtesy of Allied Materials Corp.)

through separate hoses, and intimately mixed at the nozzle. They are extruded under pressure into the joint. The machine is usually truck- or trailer-mounted and can install 20 to 25 feet of contraction joint per minute. It is essential that temperature control be maintained on both components, since temperature variation will cause uneven volumes to be pumped and will throw off the volume ratios. Figure 19.15 shows a typical trailer-mounted machine mixer.

The low-modulus silicones have been tested in highway and pedestrial joints at several test sites for several years with apparently good success. The silicone sealants are designed to take ±50% joint movement. Spells (25) describes the chemistry and the test applications and joint properties of these promising sealants for highway joints. Cost may be a deterrent to the use of these sealants, but good performance might cause a trend in this direction.

19.7.3. Hot-Poured PVC-Based Sealants

In recent years, a new material has been introduced to the hot-pour class of sealants. This material is a PVC coal tar composition that comes in several grades. The normal grade is for average pavement, while a second grade is made to meet jet fuel resistance requirements. ASTM standards have been developed to cover requirements for these materials. The materials evidently have had excellent success over the last 10 years, and a great part of the success is due to the strict adherence to design. These materials have been primarily designed for airport runways, which normally are much thicker and also have shorter spacing between joints. The materials have also been used for highway and canal joints with excellent results. There are at least three manufacturers using this patented product and they insist that joint spacing be limited to a maximum of 15 feet for new pavement and up to 25 feet for resealing of old pavement. Joint design is different and joint width is generally ½ inch with a minimum joint depth of sealant of 1 inch. All joint sealants are recessed approximately ¼ inch. Back-up material consists of a round cotton or cellulose upholstery cord or a heat-resistant premolded urethane foam. This back-up also serves as a bond-breaker to control depth of sealant.

The material is evidently a PVC powder dispersed in a coal-tar base and is pourable. Supplied in 5-gallon pails, these materials are heated to approximately 300°F, which gels the PVC and dissolves it in the coal tar. The material is pumped into properly prepared joints and upon cooling sets to a rubbery material. The material is claimed to have self-healing properties with larger movements. The composition is evidently quite simple and the raw materials have a lower cost and are therefore more acceptable than the elastomeric two-component sealants that require elaborate mixing and metering equipment and more sophisticated formulation.

The insistence of the manufacturers and installers on exact details has lead to a surprisingly high degree of successful performance, and several manufacturers have given 5 to 10-year guarantees on properly supervised installations. Gaus and Seibel (26 and 27) cover the development, properties, and installation of these materials, which are recommended for highway, airfield pavements, and jet-fuel-resistant applications as well as canal sealants.

19.7.4. *Preformed Compression Seals*

Preformed compression seals are the fastest-growing type of highway paving seal. Although some experimental sections have been made using EPDM, silicone, butyl, and GRS rubbers, virtually all the compression seals in current use are made of neoprene. Compression seals offer greater movement capability in a joint as compared to sealants, and consequently offer a greater safety margin. Also, these seals are used whenever wider joints are needed, and designs are available that will cover joints that open to 1.7 inches. This is one area where design is carefully worked out. The manufacturer insists on knowing the lowest and highest anticipated temperatures in selecting the proper compression seal. The selected seal must still have recovery pressure even at the maximum width of at least 3 pounds per square inch.

Chevron

Square

FIGURE 19.16.   Preformed seal cross-sections.

There are hundreds of different shapes of seals currently available, but the most frequently used sections are modeled after the chevron and the rectangular shape (Figure 19.16). Each has specific areas of application. The chevron shape is used for narrower joints, and a modified chevron shape is used in wider joints. Rectangular shapes are used more in bridge joints as lock seals.

The earlier failures of some compression seals were mostly due to poor compression set resistance. This has been greatly improved by upgrading reference standards and test methods. The latest standards reflect the improvements, and ASTM D-2628 and D-3542 now include a recovery requirement of 85% after 70 hours of compression at 212°F. These require-

TABLE 19.1.   Compression Seals for Road and Pavement

Style	W	H	W min.	MIW	W max	M	JD
A	7/16	3/4	0.23	0.27	0.35	0.11	1.25
B	9/16	17/16	0.24	0.31	0.45	0.21	1.50
C	5/8	17/16	0.24	0.33	0.50	0.26	1.50
D	13/16	37/32	0.30	0.42	0.65	0.35	1.50
E	1	1.5	0.37	0.52	0.80	0.43	2.00
F	1.25	43/32	0.58	0.72	1.00	0.42	2.25
G	11/8	23/16	0.73	0.93	1.30	0.57	2.37
H	2	59/32	0.70	1.05	1.7	1.0	2.75

Note: all dimensions in inches
Key: W   seal width, nominal uncompressed
       H   seal height, nominal uncompressed
      W min.   seal width, compressed at the highest anticipated temperature
      MIW   minimum installation width
      W max.   seal width at lowest anticipated temperature
      M   total seal movement capability (W max. minus W min.)
      JD   minium recommended joint depth for seal

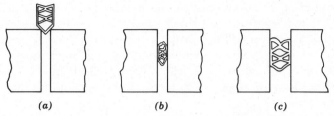

(a)   (b)   (c)

FIGURE 19.17. Preformed neoprene (a) before installation, (b) installed in a ⅜-inch joint, and (c) after joint opens to ¾ inch in cold weather.

ments should be included in all specifications for preformed compression seals.

The mathematics in preformed seals are reflected in Table 19.1, which shows the various dimensions and the capabilities of a group of road and pavement seals. Preformed seals for airfields, bridges, and other structures have different dimensions covering different sets of performance criteria. The compression seals must be properly installed in order to function properly. Figure 19.17 shows a ¹³⁄₁₆-inch-wide seal in three stages at which

FIGURE 19.18. Installation of a preformed compression seal. (Courtesy of D. S. Brown Corporation.)

it must function. The seal must not be stretched during installation. A seal that is stretched longitudinally during installation will neck down and make installation easier, but will not properly perform after it is in the joint. For example, a ⅞-inch-wide seal, when stretched, may neck down to ¾ inches. In effect, then, a ¾-inch-wide seal is being placed into the joint. In the winter when the joint opens to ¾ inches the seal could be sucked out with any air movement.

In the early days of compression sealing, the seals were placed with a flanged hand roller, which is satisfactory for small jobs. However, the roller is slow and produced some stretch in the seal. Currently several power-operated machines are available that precompress the seal, place the lubricant-adhesive on the joint wall, and insert the seal into the joint at the proper depth with a minimum of stretch. Figure 19.18 shows a typical seal installation machine; it can install 20 feet of preformed seal per minute.

Preformed seals do have a cost disadvantage. It costs $1 to $3 per foot to seal a pavement joint in new work, and the cost may go to $4 per foot or higher for resealing work. However, if preformed seals can function effectively over a 20-year period, the additional cost may be justified. Consequently, if compression seals can do the best job of protecting pavement from distress over an extended time span, they will be the highway engineer's best sealant investment.

Compression seals also have other disadvantages. In order to function properly, preformed compression seals must be installed in joints with straight, firm joint walls. Also, although the compression seals do the best job of keeping incompressible solids out of the joint, they are not watertight. Consequently, water seepage past the compression seal can corrode the load-transfer device and thus inhibit normal pavement movement. As with any other type of seal, the preformed seal depends on each slab unit to accommodate its share of the movement. Another difficulty with preformed seals is related to the forming of the joints. The seals for any given project are designed for the joint size. However, after the joints are sawed, it is often found that some of the joints have cracked through, whereas others have not. Consequently, the joints may not be uniform in size at the time of installation.

19.8.   Reference Standards

Joint sealants for highway construction are specified by highway engineers. There are approximately 75 state highway departments or turnpike authorities and over 600 city and county authorities that have some jurisdiction over pavement construction. Every state highway department has its own set of construction standards and many city and county authorities have similar codes.

The sealant standards included in most of these construction codes are modeled after applicable ASTM or federal standards. The appropriate standards have been separated into three groups. Table 19.2 covers cold-applied joint sealant standards and lists some of the earliest, now-obsolete standards for a historical background. The three federal specifications for building joint sealants are included because they are sometimes used in very limited applications. Table 19.3 covers hot-applied sealants including the latest standards on PVC-coal tar. Table 19.4 covers auxiliary materials used with highway joint sealants. Wherever possible, equivalents are shown and methods of test are given. Several ASTM standards have been discontinued and probably with time perhaps a few more will be eliminated. However, some of the materials covered by reference standards may be applicable for small jobs.

From this maze of standards, a few facts stand out. The acceptance procedure for federal specifications must be kept in mind. Manufacturers may test their own materials according to the reference standard requirements and may state in their advertisements that the material meets or exceeds a given federal specification. Fortunately, most sealant formulators are in business to stay, and value their reputations. However, it should be apparent that it would be quite simple for a fly-by-night compounder to select the recovery requirement from one standard, the bond test requirement from another, and the elongation requirement from a third standard. This compounder could advertise a pretty convincing case for a sealant that was actually useless.

Many state highway departments now have well-equipped laboratories for acceptance testing, and most states will insist on laboratory verification or actual field tests before accepting a new sealant material. Most state highway departments are quick to test new materials but very conservative about actual acceptance. City and county jurisdictions usually have competent personnel but are not equipped with testing laboratories. Consequently, they must rely on the guide specifications or leadership from state highway departments.

From the long list of standards given in the tables, several stand out as being significant. For the hot-pour asphaltic materials, ASTM D-3405, which replaces federal specification SS-S-1401, stands out as the better standard on rubberized asphalts, and ASTM D-3358, which replaces SS-S-1614, is the jet-fuel-resistant counterpart. The hot pour PVC/CT types now appear to be the best materials for airfield runways and highways. Here, ASTM D-3406 is a new standard, and ASTM D-3569 also includes jet-fuel resistance. Materials meeting both these two standards have been used with excellent results on many airfield runways and highways over the last 10 years.

In the field of elastomeric sealants, a good guide specification simply does not exist. The applicable documents, SS-S-00195 and SS-S-00200c, known as the "coal tar specs," were written around 1959. ASTM D-1852

TABLE 19.2. Cold Applied Joint Sealant Standards

Federal Spec.	ASTM Equivalents		Product Description
	Standard	Methods of Test	
SS-S-170 (obsolete)	none	none	Two-component; synthetic rubber, jet-fuel-resistant
SS-S-200C (replaced 170)	D-1852 (discontinued)	D-1853 (discontinued)	Same as above; machine mixed
SS-S-158 (obsolete)	none	none	Asphaltic; solvent type, emulsion type
SS-S-00195	D-1850	D-1851	Two-component; polymer type, machine mixed
SS-S-156 (obsolete)	none	none	Emulsion type (no composition restriction)
SS-S-159b (obsolete)	none	none	Multicomponent, mastic type
SS-S-171 (obsolete)	none	none	Asphalt, mineral-filled
TT-S-00227E	C-920	C-920	Multicomponent elastomeric sealant
TT-S-00230C	C-920	C-920	One-component elastomeric sealant
TT-S-001543a	C-920	C-920	One-component silicone sealant

TABLE 19.3. Hot-Applied Joint Sealant Standards

| Federal Spec. | ASTM Equivalents | | Product Description |
	Standard	Methods of Test	
SS-F-336 (obsolete)	none	none	One-component, hot-applied
SS-S-164 (obsolete)	D-1190	D-1191	One-component, hot-applied
SS-S-167 (obsolete)	D-1854	D-1855	One-component, hot-applied JFR
SS-S-1401 (replaces 167 and obsolete)	D-3405	D-3407	One-component, hot-applied
SS-S-1614 (obsolete)	D-3581	D-3582	One-component, hot-applied JFR
none	D-3406	D-3408	One-component, hot-applied PVC/CT
none	D-3569	D-3583	One-component, hot-applied PVC/CT JFR

TABLE 19.4. Standards on Auxiliary Materials for Highway Joints

Industry	ASTM Specifications	Product Description
	C-509	Cellular elastomeric preformed gasket
	D-1752	Sponge rubber, cork, and self-expanding cork joint fillers
	D-1751	Cork or cane fiber with felt and bituminous binders
	D-545	Cork, fiber seals with bituminous binders
AASHO-M-220	D-2628	Preformed neoprene seals for concrete pavements
	D-2835	Lubricant for installation of preformed seals
	D-3542	Preformed neoprene seals for bridges

is the ASTM counterpart of SS-S-00200c and was discontinued. It may very well be that the PVC/CT compounds have taken over in this area because of their better performance.

The standards for preformed compression seals were written around a tentative specification published by Dupont. This standard was modified somewhat by researchers in some of the highway departments and government agencies in close cooperation with industry to develop ASTM D-2628 and its industrial counterpart, AASHTO M-220.

Several of the standards list resistance to jet fuel as a requirement. These standards were obviously written for airport pavement work and have been adopted into the highway field.

19.9. Crack Sealing

Cracks occur in both asphaltic concrete and portland cement concrete. The sealing of the cracks may vary depending on the type of pavement and the size of the cracks.

Cracks in bituminous concrete may be either reflection cracks or random cracks. Reflection cracks are cracks in bituminous overlays placed on portland cement pavements. These cracks are fairly straight and correspond to the joint spacing in the pavement below. Random cracks may be caused by thermal change, frost heave, subgrade settlement, or any combination of these factors.

Properly sealing the cracks in bituminous pavements is a more difficult job than sealing the joints in rigid pavements. Because of the irregularity of the cracks, it is impossible to use a preformed compression seal. Elastomeric sealants are useless because there is no sound joint interface to which the sealant can adhere; also, most of the elastomers are incompatible with asphalt and, consequently, will not bond. Tar and asphalt have thus been the two materials used to seal these cracks. The tar is applied hot. The asphalts may be either a hot-poured or a cold-poured solvent-release type. The hot-poured asphaltic materials are the best choice. Cracks have sometimes been widened with a small routing tool to create a better bond interface, but the routing has not made any notable change in the sealant performance.

Sealing of large areas of random cracks (map cracking) has been accomplished by various types of seal coatings. The seal coatings, if properly applied to a clean pavement, may be effective for as long as 5 years.

The cracking of rigid pavements may take several forms, but the most prevalent is the transverse crack caused when the normal contraction is immobilized. Cracks in rigid pavements may be repaired structurally, or they may simply be sealed. The structural repair is accomplished with an epoxy adhesive that may be either poured or pumped into the crack. Structural repairs look good in theory, but have not been in use long enough to judge their effectiveness. Crack sealing in rigid pavements is subject to many of the same limitations that apply to bituminous pavements. The cracks are too irregular for compression seals. Also the cracks are characterized by aggregate pull-out and weak interfaces, and thus elastomers cannot be used. Consequently, tars and asphalts are used as with the bituminous pavements.

19.10.   *Sealants in Airport Paving*

The problem of sealing airport pavement joints is somewhat different from that of sealing highway pavement. This difference in sealing practice is mainly due to the difference in slab depth and panel size. In highway construction, the average slab thickness is approximately 9 inches. In airport construction, the slab thickness may be twice this amount or even more, because of the heavier wheel loads the pavement has to support. Consequently, airport pavement is slower to respond to temperature changes than the highway slab. This difference in slab thickness also demands a different type of joint construction. The ordinary saw-cut contraction joint is usually not applicable to airport paving.

In highway construction using a contraction joint spacing of 50 feet, the highway is generally paved in a two-lane width, and then the joints are sawed. Contraction joints are placed every 50 feet, and a longitudinal joint is sawed that divides the pavement into two 12-foot lanes. The pavement slab is thus subdivided into panels 50 feet long by 12 feet wide. In airport construction, especially in parking and ramp areas, the pavement is usually placed in square panels approximately 25 feet on each side. There is no real differentiation between longitudinal and transverse joints. Airport runways, of course, are several thousand feet long and about 100 feet wide, so that transverse and longitudinal joints can be identified; however, the construction is still usually in the form of square panels.

Because of the relatively short joint spacing and greater slab thickness, there is little or no curl in airport pavements. Almost all of the movement that occurs in airport pavements can be attributed to temperature and moisture changes.

Joints in airport pavements are generally formed with some form of insert. The insert may be a plate or it may be a sawed-out fiberboard insert. The sawed-out insert has become the more popular method of joint forming.

Materials used in airport pavements are similar to those used in highway construction, except for the addition of a requirement for jet-fuel resistance. The sealant and its adhesive bond to the concrete pavement must not be affected by jet-fuel spillage. This is especially critical in parking and ramp areas where aircraft are serviced. ASTM D-3406 is the best standard to use for this area of application. If jet-fuel resistance is required, then D-3569

FIGURE 19.19. Installation of PVC-coal-tar sealant into airport runways. (a) Tractor with rear-mounted hydraulic joint plow designed to follow joint; (b) Cutting tool used in the plowing of joints, preferably straight-sided without a taper. (Courtesy Superior Products Co., Inc.) *(continued)*

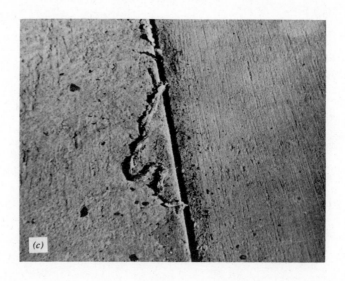

FIGURE 19.19. (c) Joint after treatment by cutting tool. All old sealant loosened from side walls; (d) Sandblasting of joint sidewalls. (Courtesy Superior Products Co., Inc.) *(continued)*

FIGURE 19.19. (e) Power sweeper used in cleanup of old sealant and blasted sand from pavement surface; (f) The blowing clean of a joint with 100 psi compressed air just prior to bond breaker installation. (Courtesy Superior Products Co., Inc.) *(continued)*

is the answer. Preformed compression seals have been also used in airport work, but their growth in this area may be slower than in highway sealing where the movement is greater. Figure 19.19 shows the complete sequence of resealing with a PVC/CT joint sealant.

Actually the construction of airports can be divided into two major areas: civilian and military construction. In military construction, the federal specifications apply directly. In civilian airport construction, the work is subject to the jurisdiction of the Federal Aviation Authority. The FAA has its own set of standards and recommended practices, but the sealant stand-

FIGURE 19.19. (g) Saw cutting of new joints or widening old joints. Diamond-tipped blades with water for cooling and lubrication; (h) Insertion of back-up rod and bond breaker into clean joint just prior to sealing. (Courtesy Superior Products Co., Inc.) *(continued)*

FIGURE 19.19. (i) Proper sealing operation, sealing joint with sealing tip at bottom of joint, filling joints ¼ inch below flush from the pavement surface; (j) Completed joint, sealed below flush). (Courtesy Superior Products Co., Inc.)

ards are, for the most part, modeled after the federal specifications. In recent years, the General Services Administration has relinquished its authority in some areas and has chosen many ASTM standards as equivalents for its own specifications.

One additional sealant requirement is brought about by the difference between civilian and military aircraft. On almost all civilian aircraft, the jet engines are horizontal, so that there is usually no direct blast of jet exhaust down against the pavement. Many military aircraft, on the other hand have the jet engines mounted at a slight angle of incidence, so that there is a cone of jet blast directed toward the pavement behind the aircraft. Sealants selected for use in these areas should be selected to withstand this jet blast.

20

Sealants in Bridge Expansion Joints

20.1. Introduction

Bridges are harder to seal properly than pavement joints. This is because bridges are subject to all the forces that move pavements, and a few more besides. In addition to being harder to seal, bridge joints are more critical than pavement joints. Poorly sealed bridge joints, as well as being very expensive to repair, can result in conditions that are hazardous to the motorist.

For drainage purposes, the bridge is usually on at least a minimum grade and may also be crowned. Consequently, the unsealed or poorly sealed expansion joint simply serves as a funnel to direct the flow of water and deicer solutions onto the bridge seals and pier caps. In ordinary pavement joints, incompressibles in the joint are the major problem; however in bridge joints the most serious problem is water leakage. Materials for use as bridge sealants include elastomeric cold-pour sealants, compression seals, strip seals, modular expansion joint seals, and other products.

Joints and bearings are often the forgotten structural design in bridges, but there has been considerable interest in them in recent years. A recent conference held in Niagara Falls, New York was organized by the American Concrete Institute in 1980 and over 60 papers were presented by authors from 11 countries. The purpose of the conference was to present the state of the art in joint sealing and bearings for concrete structures, and to provide a forum for discussion and the exchange of ideas and point the direction for future developments. Reference has been made to several contributors, and anyone interested in the latest technology on materials for bridges and highways should obtain Publication SP-70 from the American Concrete Institute.

20.2. History

In the past, bridge joints have been handled by a wide variety of methods. Sliding plates, finger and toothed joints were in use in the 1930s in order to bridge the open joint where movement was in excess of the available sealants. Many states used gutters and downspouts to carry off drainage water. The first sealing materials to be used were the hot-poured asphalts. In the 1950s polysulfides were first used as bridge-joint sealants and in the early 1960s preformed neoprene seals were introduced. Bridge joints are usually much wider than pavement joints and the preformed seals were therefore developed for this application. Curable sealants still find use for joints up to ¾ inches wide where the joint movement does not exceed ±25%. Compression seals are satisfactory for joints having up to 2 and 4 inches of movement, with strip seals and box seals used in slightly wider joints but for different applications.

20.3. Why Seal the Joints?

Expansion joints in bridges are sealed primarily to keep out water. The deicing salts used on the bridges and pavements in winter are excel-

FIGURE 20.1.   Pier cap that has begun to deteriorate due to deicing chemicals.

FIGURE 20.2. Deterioration of a bridge column.

lent for their purpose, but are destructive to concrete. Water and deicer solutions passing through the bridge joints corrode both the ends of steel beams and the roller bearings on which the bridge rests. This water also carries dirt and debris with it that in addition to staining pier caps and columns, may inhibit the normal action of roller or sliding bearings. Water washing down through the joint and over the bridge seat also contributes to the migration of back slopes. In colder climates, water and icicles may drop onto the roadway and traffic below. The most severe danger, however, is the heavy flow and concentration of deicer solutions onto bridge seats and pier caps. These are shaded areas and the deicers have ample time to saturate the concrete for maximum destructive effect. All in all, many state bridge engineers are in agreement that it is easier to try to seal the joint than to maintain an open joint.

Some of the results of poor joint sealing are shown in Figures 20.1 and 20.2. Structural members subject to attack have deteriorated in a remarkably short time. Figure 20.1 shows a pier cap that has begun to deteriorate. Figure 20.2 shows a badly disintegrated bridge column which was approximately 10 years old when the photograph was taken.

20.4.   Structure of the Joint

Bridge expansion joints resemble isolation joints in buildings. They are "through the structure" joints that completely separate the bridge from the approach pavement. The structure of the joint will vary depending on the span length and type of construction. Figure 20.3 shows a sealant joint that can be used on very short spans and places the joint seal into a butt joint, so that the seal is subjected to tension–compression movement cycles. Unless the movements are restricted to a maximum of ±25% for a good elastomeric sealant, the sealant will fail; and generally with the additional vibrations that can occur, sealants have a shorter life. Perhaps this might be a place for the lower-modulus silicone sealants, but this remains to be proved.

Figure 20.4 shows a sliding-plate joint that is also applicable to short spans. There has been an improvement in this system by using strip seals under the plates to prevent the water from passing through the joint.

The construction of the finger-plate joint shown in Figure 20.5 was designed to give traffic support in wide moving joints. These are open joints which again introduce the problems of water corrosion and have been in part replaced by closed sealing systems. Figure 20.6 shows a steel reinforced preformed elastomeric seal that is one of many designs that have been introduced in recent years.

The extruded neoprene seals for bridges are quite varied, and Figure 20.7 shows examples of typical compression and lock-in compression seals such as box seals and strip seals.

Compression seals have working limits that vary from 70 to 80% of nominal width to approximately 40 to 50% of nominal width, which are the

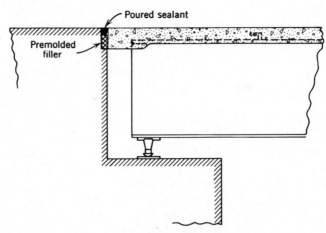

FIGURE 20.3.   Expansion joint in a short-span steel-beam bridge.

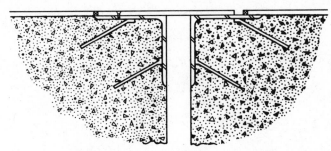

FIGURE 20.4. Sliding-plate joint.

extremes in the expected movement. Thus a 4-inch-wide compression seal would work satisfactorily in a joint with width dimensions varying from approximately 1.75 inches to 3.4 inches during the diurnal and annual movements. The compression seals require high recovery and maintain a seal by virtue of always being in compression and maintaining at least 3 pounds per square inch pressure against the sides at all times.

The lock-in compression seals have greater capability and do not function by maintaining compression set resistance. They simply move with the bridge and are locked into the sides of the joint.

Box seals are generally used in engineered expansion joint systems containing 2 to 10 box seals. A typical box seal would have an overall width of 4.5 inches and a height of 3.5 inches; when locked in place it would have a maximum movement capability of 3 inches where the opening would vary from 0.06 inches to 3.06 inches. Thus a box seal would be used for every 3 inches of expected movement. Using a hypothetical median of 1.9

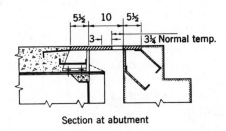

Section at abutment

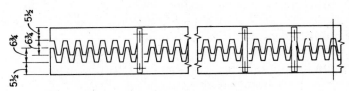

FIGURE 20.5. Finger-plate expansion joint.

FIGURE 20.6.   Steel reinforced, preformed elastomeric seal. (Courtesy of General Tire and Rubber Co.)

inches on the seal, this imparts approximately ±50% joint movement on the seal.

Strip seals are lower-cost systems than box seals and work effectively in joints with movement up to 4 inches. Thus a strip seal having a width of 5.6 inches overall could take normal movements of 4 inches, where the actual opening may vary from ¼ inch to 4¼ inches; and in abnormal use the opening would vary from ⅛ inches to almost 6 inches and the strip seal would actually be stretched. The strip seal expansion joint systems are ideally suited to reconstruction of joints in existing structures. They can be simply field welded over finger-slider plates, and other steel faced joints, or can be directly anchored to asphalt and concrete overlays. Figure 20.12 shows a typical diagram of strip seals during movement. This is only one of many types of strip seals available. Heavier-duty seals require anchorage between strip seal glands and the structure.

The smaller compression seals can be placed directly into straight-sided concrete joints. For all other types of seals, specially designed steel extrusions must be used to properly hold the box seals or the various types of strip seals. Proper design calls for anchorage bolts for the various steel extrusions to stand up under continuous heavy traffic.

Engineered expansion joint systems are made up using 2 to 10 box seals capable of handling 6 to 30 inches of movement, respectively. Figure 20.8 illustrates a modular joint system with 4 box seals.

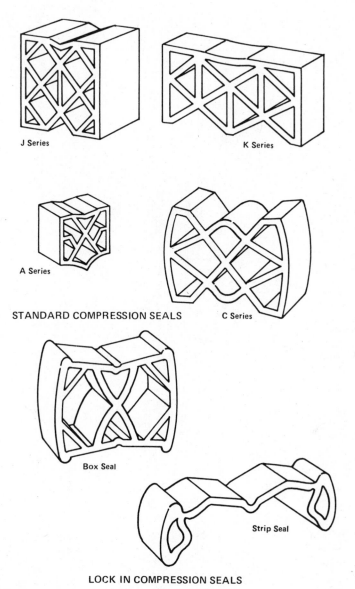

J Series

K Series

A Series

STANDARD COMPRESSION SEALS

C Series

Box Seal

Strip Seal

LOCK IN COMPRESSION SEALS

FIGURE 20.7. Examples of compression seals, box seals, and strip seals. (Courtesy of Acme Highway Products Co.)

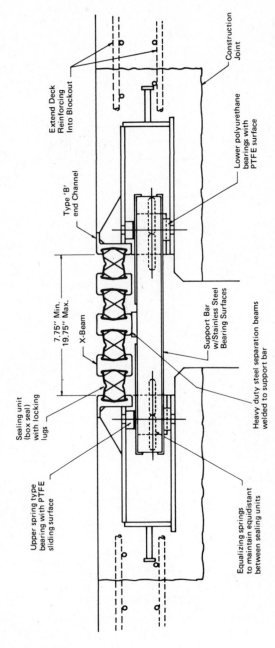

Upper spring type bearing with PTFE sliding surface

Sealing unit (box seal) with locking lugs

Extend Deck Reinforcing Into Blockout

Type 'B' end Channel

7.75" Min. 19.75" Max.

X-Beam

Construction Joint

Lower polyurethane bearings with PTFE surface

Support Bar w/Stainless Steel Bearing Surfaces

Heavy duty steel separation beams welded to support bar

Equalizing springs to maintain equidistant between sealing units

FIGURE 20.8. Engineered modular expansion joint system with many improvements. (Courtesy of Acme Highway Products Co.)

252

20.5. How Bridges Move

The movement of bridges is assumed to be due to thermal expansion and contraction. In her studies of a number of concrete bridges, Emerson (28) found other factors that affected concrete movement such as drying, shrinkage, and creep, which are irreversible and occur during the first 6 months after construction. The author also found that the values of the coefficient of linear expansion varied depending on the value of the coefficient for the aggregate. Emerson studied many bridges including concrete box, (single and multicell) concrete beam, and slab and solid concrete deck bridges, and found peculiarities that were unique to each bridge.

The deflection of the bridge under live load is also responsible for joint movement. The midspan deflection of a bridge under traffic causes an end rotation of the structure which opens the expansion joint. Simple computations show that a 60-foot steel beam bridge with a concrete deck can cause the expansion joint to move by an amount equal to one-half the yearly thermal movement of the bridge every time a 20-ton truck passes over the structure. These computations have been checked by the electronic instrumentation of many structures. In winter months, when the bridge contracts, this movement should be added to the movement caused by contraction of the structure. The trend in bridge design is toward larger, more flexible structures, and thus deflection movements cannot be ignored.

The braking action of traffic on bridges is also responsible for bridge movement on short-span structures. In a new bridge, this longitudinal thrust of the bridge puts a dynamic load of tension on the seal at one end, and compression at the other end. The actual amount of this movement is unknown because it is so difficult to separate it from other bridge movements, even with electronic instrumentation. It is known, however, that as the bridge ages, there is a longitudinal shoving of the structure in the direction of traffic. This factor alone could necessitate the resealing of a structure after a period of a few years.

Continuous construction has also served to increase joint movements. A three-span bridge with spans of 50, 80, and 50 feet might be constructed

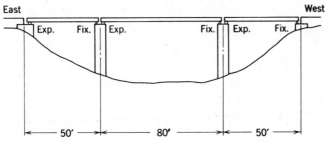

FIGURE 20.8a.   Three-span bridge built as simple spans.

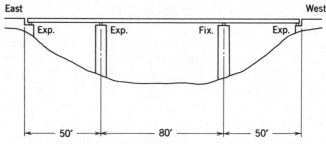

FIGURE 20.9.   Three-span continuous bridge.

as three simple spans with a fixed bearing and an expansion bearing on each pier cap (Figure 20.8a). However, many bridges in this same span range are now built as three-span continuous structures. With this type of construction, the support bearings might be arranged as shown in Figure 20.9. This structure has only one fixed bearing and, consequently, the expansion joint at the east end of the structure must accommodate a total movement due to 80 plus 50 (or 130) feet of span length. In the simple beam bridge of Figure 20.8, the movements for each span are accommodated at the ends of the individual spans.

Bridges are also known to be more responsive to thermal changes than are pavements on grade. Highway pavement surface is subject to great thermal changes, but the bottom of the pavement slab is fairly well insulated from these thermal fluctuations. A bridge structure, on the other hand, is exposed to temperature changes under the bridge as well as at the top surface.

Another factor that must be considered in the design of bridge joints is the skew of the bridge structure. By far the larger number of bridges are installed at some angle of skew. Figure 20.10 shows a plan view of a typical skew bridge. In the skew bridge, the main structure members are aligned with the direction of traffic, so that the thermal expansion and contraction is in this direction. The effect of the skew is to place the joint seal in longitudinal shear as well as tension and compression. This combination stress makes the joint very difficult to seal. Preformed seals tend to bunch up and "walk" out of the joint. Strip seals have the greater movement capability

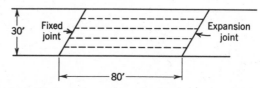

FIGURE 20.10.   Plan view of a single-span skew bridge.

and give the least resistance to rocking movements of a skew joint. The seals can easily function if joint faces are uneven, or not parallel, and can even take considerable rocking movement in the joint sides due to traffic movement. Poured sealants are subject to a peeling action that tends to break down the adhesive bond of the sealants to the concrete.

20.6. Materials Used

The hot-poured asphalt sealants are not used to any great extent due to the dynamic movement of bridges, but are used somewhat in very short span bridges. These materials do not have enough resiliency and elongation capacity to cope with the movement of bridge structures.

The elastomeric sealants, when first introduced, were used in a great many bridge joints without much success. Since further research has shown the extent of bridge movements, the manufacturers of elastomeric sealants have confined their efforts to the shorter span structures and have fared much better.

The elastomeric sealants for use in bridge joints should have the same properties as the pavement sealants. Although incompressibles are not the most serious problem in bridge joints, the sealant should have a Shore A hardness of 15 to 25 to prevent puncture by sharp stones and grit. The joint should in most cases not exceed more than ¾ inches in width, and the sealant should not take more than ±25% joint movement or ±³⁄₁₆ inches movement at any time during the life of the sealant. It is imperative that the temperature range be determined since the bridges will undoubtedly reflect the extremes in air temperature for the location and also include an additional safety factor to cover deflection due to traffic loading and travel. Temperature gradients of 120 to 140°F are normal for many parts of the U.S. and could be greater depending on heat absorption. The sealant must have good adhesion to both concrete and steel and also exhibit recovery of at least 75% after being blocked in a strained position at 50% compression for 24 hours at an elevated temperature of 158°F. Sealants based on urethane and silicone will meet this requirement. The 2:1 width to depth ratio should be used with proper back-up bond breakers and joint cleaning. California has approved the use of urethane sealants for use on short-span bridges with small movement, and the use of sealants in these areas will probably work satisfactorily if proper precautions are given to movement consideration.

Preformed seals based on neoprene include compression seals, box seals, and lock strip seals, and have virtually taken over the entire sealing market of medium- and long-span structures. Although other rubbers have been evaluated, neoprene has been time proven and other materials must do likewise before ultimate acceptance.

Compression seals are limited to the amount of movement that they can take. For road, pavement, and airfield seals, the maximum amount of movement is on the order of 1¾ inches, in which case the seals are up to 2 inches in width before insertion. These seals are generally recessed ¼ inch below the surface to protect the seal from abrasion. Also, the seals should be placed into a designed groove so that the seal will be seated on shoulders. There are two types of lubricant-adhesive currently in use. The first is based on a neoprene adhesive at approximately 25% solids content and is used in concrete joints where the adhesive primes the joint faces and seals small holes and imperfections in concrete surfaces. The adhesive is also resistant to water and prevents the seal from working out of the joint with vibration and expansion movements. The second is a one-component moisture-curing urethane adhesive at approximately 75% solids content and is used with steel and other metal-faced joints. Adhesive-lubricants should provide excellent adhesion of the neoprene to the surfaces involved and should not be affected by water, deicing solutions, temperature variation, or the weather. Standards for lubricant-adhesives are presently being balloted by ASTM.

Compression seals for bridges are heavier construction extrusions with heavier walls and can be up to 6 inches in width, which will take up to 2.85 inches of movement. The selection of the size of the compression seal is determined by calculating the maximum movement and then allowing 15% deviation for concrete shrinkage, creep, and moisture rating. Once the joint movement is determined, the seal that has a slightly higher movement capability is selected, and this seal will have a minimum and maximum width characteristic that will be incorporated into the joint design. In other words, the joint width of the facing steel elements will have to be positioned so that the seal can work within the limits set.

Compression seals are covered by ASTM C-2628, which has very stringent recovery requirements added in recent years to cover the necessary recovery requirements. The toughest part of the standard is a requirement of 85% recovery after 70 hours of deflection at 50% at 212°F minimum. Certification from an independant laboratory should be required for conformance to this standard.

Lock seals do not perform on a compression basis. They have been designed so that the curvature of the seals fit the curvatures of the steel cavity, and the side walls of the seal are found to coincide with the contour of the steel cavity, thereby creating a watertight seal. The seal webs are designed so that they fold within themselves and the seal is capable of greater movement than compression seals of similar nominal width. The lock seal in normal and closed position is shown in Figure 20.11. The lock seal comes in essentially one size and is used singly or in multiples to take care of a movement increment of 3 inches per seal. The steel extrusion comes in various designs to accommodate methods of attachment. The engineered expansion joint is completely assembled at the plant and delivered intact to the job site. The width of the expansion joint is preset at the factory to

Seal in a relaxed state.

**Seal compressed
to design limit.**

FIGURE 20.11. Compression stages of box seal elements. (Courtesy of Acme Highway Products Corp.)

the required setting; however, if the ambient temperature at the time of placement varies more than ±5°F from the preset temperature, the width must be readjusted at the site. Generally there are two methods of attachment. One is to place the unit in position once the superstructure supporting members have been erected, and then complete the deck concrete. The second method is to leave a recess in the concrete deck to receive the unit at a later date, and then complete the work. Temporary erection beams, lifting hooks, width-adjusting devices, and the like, are normally provided so that the unit can be safely handled and accurately installed.

There are many designs for strip seals depending on available methods of anchorage. Strip seals do not depend on optimum resilience for satisfactory performance. The seals are designed for easy reconstruction of existing joints on structures. The extruded metal interfaces can be simply field-welded over finger-slider plates, and other steel-faced joints, or can be directly anchored into a concrete deck prior to asphalt or concrete overlaying. Strip seals are capable of movements up to 4 inches. Figure 20.12 shows one style of strip seal in various configurations to demonstrate movement capability.

Other types of seals based on the strip-seal principle are made of combinations of metal retainers or steel-reinforced neoprene retainers that mechanically lock the seal in place. The retainers lock the seals in place so that they cannot be pulled out by traffic or skew forces, and act to distribute the sealing capability along the entire length of the gland. These seals can take up to 4 inches in movement and because of their locked-in mechanisms are capable of greater skew considerations. However, if the skew angles are large, then this will affect the maximum opening to allow for distortion in the seal. The seals are sold with epoxy resins for anchoring bolts in the concrete and urethane sealants to fill cavities for bolt holes. One of the

Diagram of Mounting Unit

The mounting unit is a one piece self contained locking assembly.

The unit consists of upper and lower retainer lugs plus an inner and outer wall. See diagram.

⓿ Upper Lug
❷ Lower Lug

❸ Outer Wall
❹ Inner Wall
❺ Strip Gland

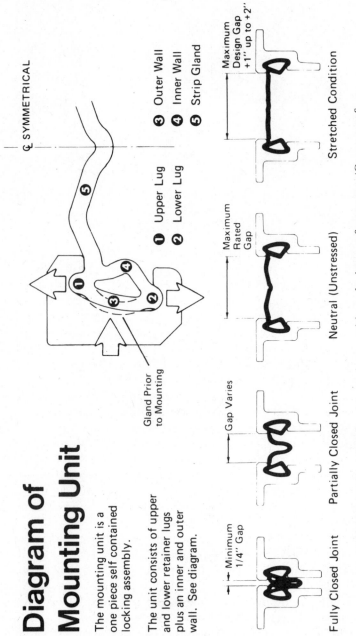

FIGURE 20.12. Strip seal in various positions during performance. (Courtesy of Acme Highway Products Corp.)

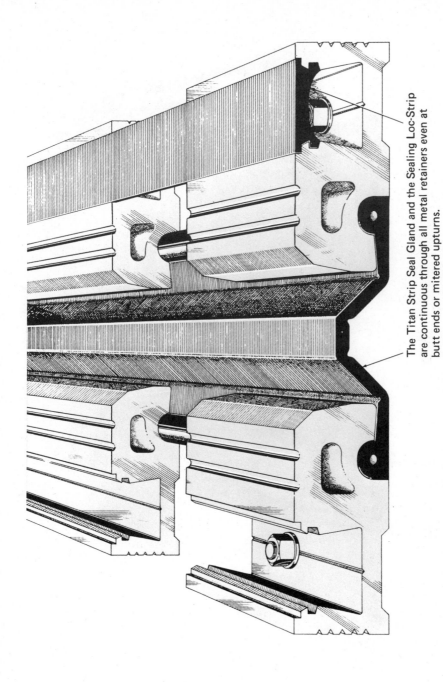

The Titan Strip Seal Gland and the Sealing Loc-Strip are continuous through all metal retainers even at butt ends or mitered upturns.

FIGURE 20.13. Strip seal gland with metal retainers and sealing lock strip. (Courtesy of Acme Highway Products Corp.)

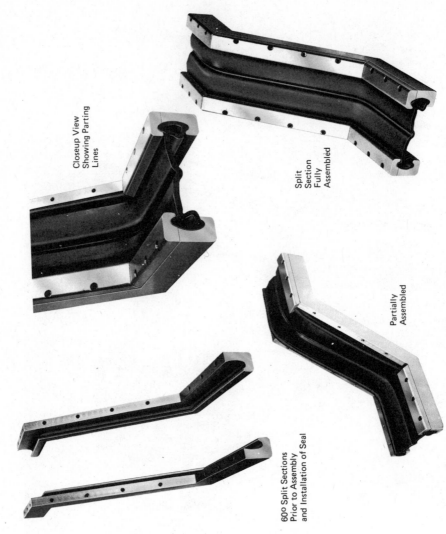

Closeup View
Showing Parting
Lines

Split
Section
Fully
Assembled

Partially
Assembled

60° Split Sections
Prior to Assembly
and Installation of Seal

FIGURE 20.14. Strip seal at curb showing split-section mechanicals. (Courtesy of Acme Highway Products Corp.)

principle advantages of these seals is the low profile and easier anchorage to flat decks. Figure 20.13 illustrates one type of strip-seal system based on a strip-seal gland, metal retainers, and a sealing lock-strip. Strip seals, because they lack webs, are easily bent around corners, and Figure 20.14 shows a strip seal at a curb that can be carried up a wall if necessary. Puccio (29) and Watson (30) have covered additional factors for consideration in bridge and highway joint design, and offer a complete line of compression seals, box seals, and various strip seals to cover a multitude of problems.

20.7. Reference Standards

The two existing ASTM standards that cover neoprene seals for pavement and bridge joints are D-2628, "Preformed Polychloroprene Elastomeric Joint Seals for Concrete Pavement," and D-3542, "Preformed Polychloroprene Elastomeric Joint Seals for Bridges." Both standards are almost identical with respect to requirements, and both require certification by an outside laboratory. D-3542 differs in that it has additional compression–deflection requirements. Both standards have stringent recovery requirements that include 85% recovery after 70 hours of compression at 50% deflection at 212°F. The recovery requirement is waived where strip seals are used, since they do not operate by compression.

21

Adhesives

21.1. Introduction

Adhesives are here to stay. History shows us that adhesives have been used since the time of the early Egyptians and Assyrians. The Egyptians used adhesives to bond papyrus, and pieces of veneered furniture have been found in the tombs of some of the pharaohs. These adhesives were undoubtedly of animal origin. The Assyrians not only contributed mud brick, our earliest known molded building material, but in many cases bonded the brick together with a cementitious mortar based on clay. These same Assyrians discovered bitumen springs near the Euphrates River and used this pitch as a cementing material. The Greeks were the pioneers of faced and veneered construction. They often built structures of limestone that were veneered with a fine-grained marble. The marble was sometimes attached with a mortar adhesive. The early Romans also used a faced concrete type of construction. The Chinese used mortars in their structural work and animal glues in their decorative building work.

In the Western world, mortars, bitumens, and starches have been used freely, but the growth in the use of adhesives up until the 1930s has been pretty much the growth in the use of animal glues. Within the last five decades an explosion has taken place in the number and types of materials available for adhesive bonding: epoxies, polyesters, phenols, polyvinyl acetates, anaerobics—the list seems almost endless and is still growing.

It looks almost impossible to discuss in detail all the adhesives in use in the construction industry today. However, this is not really necessary. On a value basis, approximately 60% of the adhesives manufactured are for captive consumption, as when the company that manufactures the adhesive also produces the finished fiberboard. Captive adhesives will be touched on in this book, but the emphasis will be placed on those materials the architect or engineer may specify for use in construction.

TABLE 21.1. Adhesives in Construction

Type	Composition	Volume (Millions of pounds)	Usage
Roofing cements	asphalt and coal tar pitch, neoprene	over 500	specified by architect
Floor and wall tile adhesives	asphalt, rubber, polyvinyl acetate	over 500	specified by architect
Thermal insulation binders	phenolic resins	over 300	captive
Concrete adhesives	epoxy resins, polyesters	over 35	specified by engineer
Plywood—particle board and glued laminated members	urea resin, casein, resorcinol, phenolic	over 500	captive
Wallpaper pastes	starch	over 300	specified by architect
Pressure-sensitive adhesive tapes	rubber, starch, PVA, cellulosics	unknown	specified by architect
Paper	PVA, vinyl acetate, starch, cellulosics	over 500	captive
Miscellaneous	anaerobic, cyanoacrylates, etc.	over 2	over-the-counter

TABLE 21.2. Selecting Adhesives

	Leather	Paper	Wood	Felt	Fabrics	Vinyl plastics	Phenolic plastics	Rubber	Tile, etc.	Masonite	Glass	Metals
Metals	1, 4, 21, 24, 25	1, 21, 22	1, 4, 11, 13, 21, 31, 32, 33, 35, 36	1, 5, 22	1, 21, 22, 24	25, 26	3, 13, 21, 31, 32, 33, 35, 36	13, 21, 22, 31, 32, 33, 35, 36	5, 6, 13, 22, 35, 36	5, 6, 13, 22	13, 32, 33, 34, 35	11, 13, 31, 32, 33, 36
Glass, ceramics	1, 4, 13, 24	1, 21, 22	1, 13, 21, 31, 32, 33, 35, 36	1, 5, 6, 21, 22	1, 21, 22, 24	25, 36	3, 13, 21, 31, 35, 36	21, 22, 31, 35, 36	4, 22		4, 13, 32	
Tile, etc.	1, 4, 21, 24	1, 21, 22	1, 5, 6, 21, 22	5, 6, 21, 22	5, 6, 21, 22, 24	25, 36	3, 13, 36	21, 22, 31, 35, 36	4, 5, 6, 22	5, 8, 13		
Masonite	1, 21, 24	1, 21, 22	1, 5, 6, 21, 22	5, 6, 21, 22	5, 6, 21, 22, 24	25, 36	3, 13, 36	21, 22, 31, 35, 36	5, 6, 22			
Rubber	21, 24	21, 22	21, 22, 33, 35, 36	21, 22	21, 22, 23	25, 36	21, 22, 36	21, 22, 31, 35, 36				
Phenolic plastics	21, 24, 25	21, 22	11, 13, 21, 24, 32, 33, 36	21, 22, 25, 36	21, 22, 24, 25	36	13, 32, 33, 36					

	Vinyl plastics	Fabrics	Felt	Wood	Paper	Leather
Vinyl plastics	21					
Fabrics	21, 22, 23, 24	21, 22, 23	21	21	21	25, 36
Felt	21, 22, 23, 24	21, 22, 23	5, 21, 22, 23	5, 22	1, 21, 22, 23	
Wood	21, 22, 23, 24	2, 21, 22	1, 11, 12, 14, 15, 16			
Paper	21, 22, 23, 24	2, 4, 21				
Leather	1, 4, 21, 22, 23, 24					

Adhesive number code:

Thermoplastic
(1) Polyvinyl acetate
(2) Polyvinyl alcohol
(3) Acrylic
(4) Cellulose nitrate
(5) Asphalt
(6) Olcoresin

Thermosetting
(11) Phenol formaldehyde (phenolic)
(12) Resorcinol, phenol-resorcinol
(13) Epoxy
(14) Urea formaldehyde
(15) Melamine, melamine-urea formaldehyde
(16) Alkyd

Elastomeric
(21) Natural rubber
(22) Reclaim rubber
(23) Butadiene-styrene rubber
(24) Neoprene
(25) Buna-N
(26) Silicone

Resin Blends
(31) Phenolic-vinyl
(32) Phenolic-polyvinyl butyral
(33) Phenolic-polyvinyl formal
(34) Phenolic-nylon
(35) Phenolic-neoprene
(36) Phenolic-butadiene-acrylonitrile rubber

21.2.   The Market

Accurate estimates of adhesive consumption by the construction industry are hard to obtain because these materials are sold under such names as glues, mortars, adhesives, mucilages, and resins. It has been estimated that over 300 companies manufacture and market adhesives under brand names. A similar number of companies manufacture adhesives solely for captive use. Some of the larger companies channel their output to both captive and outside uses, but this is fairly rare. The picture that emerges is of an industry dominated not by a few large producers but rather by a cluster of smaller companies all competing for a share of the market. Plants tend to be small and widely scattered geographically to serve specific outlets. Even the larger companies tend to have smaller plants, scattered to give good geographical coverage. The effect of this industry distribution on the construction business is that quality control on an industry-wide basis is hard to achieve. Also, standards are harder to write and enforce than in other areas, such as in steel construction.

Table 21.1 gives an estimate of the major uses of adhesives in the construction industry. This table does not attempt to list all the adhesives used for the various applications. The compositions shown are simply representative of a given group and may not be the best adhesives available. Subsequent chapters will list advantages and disadvantages of the various adhesives within each group.

Table 21.2, derived by Reinhart and Callomon (31), gives an idea of the type of adhesives that could be used to adhere various materials together. This table is not up-to-date but does list 36 types of adhesives that were available in the late 1950s and that have been added to in recent years.

More and more uses are being found every day for adhesives in construction. Adhesives for concrete are growing rapidly, and so are the pressure-sensitive tapes. Metal-to-metal bonding is now being used with nonstructural components. This widening of the field of application of adhesive bonding has been due to the growth of the thermosetting resins, such as epoxies and polyesters. These synthetic resin adhesives are often desirable because they behave better in moist environments than do animal or vegetable glues. Recent additions to adhesives are the anaerobic adhesives which set in the absence of oxygen, and the cyanoacrylates which are the current rage as very-rapid-setting super-adhesives that will literally stick anything to anything. They are expensive over-the-counter items, as are many other glues and adhesives.

21.2.1.   Types of Adhesives

Because of the large number of adhesives presently available both on the market and for captive use, a majority of the various adhesives will be lightly covered as to chemical type, chemistry, volume, and application areas. It is hoped that this brief treatment will answer some of the questions pertaining to uses that continually crop up. For additional information, Skeist (32) has assembled an excellent *Handbook on Adhesives* that goes into great detail on the chemistry, properties, and uses for at least 40 major adhesives, along with an excellent treatment on adherends and bonding technology.

Inorganic adhesives include the wide range of materials described in the following paragraphs.

The *soluble silicates* are used as adhesives for metal, wood, paper wallboard, and fiber drums. Silicate cements are used for mortar. Calcium silicates used in combination with other silicates comprise the compounds used in making the various cements such as portland cement and lime cement. Lime mortars are used for plaster and setting tiles.

Animal glues are made from hides and bones that supply the collagen that is the base component in the glue. The glues are used in woodworking, gummed labels, coated abrasives, and sizing for paper and fabric.

Fish glues are made from fish skins which also supply collagen. The glues are used for adhesives for glass, ceramics, metal, wood, cork, paper, and leather.

Cassein glues are made from skim milk. These glues are used for wood laminates, composite wood beams, plywood, and paper bonding.

Soybean glues are derived from soybeans. They are also used for making plywood.

Blood glues are made from dried blood of hog and beef. They are used in combination with other materials and also for making plywood.

Starch-based glues are made from corn, wheat, potatoes, tapioca, and other starch produce. They find use in making paper, corrugated board, and paper bags. They are also used for textiles to improve their resistance to abrasion during the weaving process. This constitutes one of the largest volumes for glue.

Cellulosics are made from cotton fibers and wood pulp. The cellulose is chemically altered to form various organic derivatives such as cellulose, nitro cellulose, cellulose acetate, ethyl and methyl cellulose, and others. They have many uses. Cellulose nitrate is used in many small areas but mostly used for china, wood, metal, glass, paper, and leather. Methyl cellulose is used for wallpaper, paper making, textiles, and many other areas of application.

Natural rubber is used in latex form and allowed to dry to form the self-adhesive envelope and other touch-adhesive applications such as automobile trim.

Butyl rubber adhesives are used in many hot-melt adhesives along with resins and vinyl compounds. The hot melts are used in making insulating glass. Butyl rubber is used in tapes and pressure-sensitive tapes.

Nitrile rubber adhesives are extremely widely used. Some adhesives include various types of rubbers and resins. They are used as leather adhesives, aircraft bonding, automotive brake lining, printed circuits, and roofing. This group is extremely versatile.

SBR adhesives modified with resins are also used for pressure-sensitive tapes. They are also used with tiles, packaging, fabrics, paper, and in many other areas.

Neoprene adhesives are one of the more important groups of adhesives. They are widely used in making shoes, bonding vinyl trim in automobiles, kitchen paneling, gypsum board, and also sold over the counter in many areas of application.

Phenolic resin adhesives are used for bonding abrasives, abrasive wheels, asbestos lining, binder in the manufacture of steel molds and cores, wood bonding, laminating, contact adhesives, and many other areas.

Resorcinol adhesives are used in making water-resistant adhesives and are widely employed to make exterior-grade plywood.

Epoxy resin adhesives are used for bonding metals, glass, plastics, and wood, and are widely used for concrete and masonry repair on highways and bridges.

Polyurethane and isocyanate-based adhesives are used in many specified areas of application where outstanding performance is required: shoes, magnetic tape, furniture, automobile, textiles, garments, and packaging.

Polyvinyl acetate and polyvinyl alcohol adhesives are used to make "white glue" for home use. These glues are relatively low cost and can be used for book binding, paper bags, automobile upholstery, leather binding, and tile cements.

Vinyl acetate emulsions are widely used with cellulosics in packaging and wood adhesives. They are also used to supply adhesion to metal foils, polymer substrates, and glass.

Polyvinyl acetal is widely used to make safety glass and also hot-melt adhesives for paper, metal foil, and plastic film.

Acrylic adhesives are generally used as binders for making paper coatings, paints, varnishes, non-wovens, pigment finishes for leather, and flocking adhesives. They are used for bonding linoleum, wallpaper, ceramic tiles, pressure-sensitive tapes, and many other items.

Anaerobic adhesives become active in the absence of oxygen and form excellent bonds. They are used for locking thread assemblies, sealing threaded porous and flanged assemblies, strengthening cylindrical assemblies, and structurally bonding metal components and plastic to metal components.

Cyanoacrylate adhesives are the new so-called miracle adhesives sold over the counter to stick anything to anything. They will adhere almost any two surfaces whether metal, plastic, rubber, or wood.

Polyamide adhesives are generally used in hot-melt and thermosetting applications for aircraft, automobile, electrical, packaging, product assembly, shoe, and woodworking applications.

Silicone adhesives are used for ceramics, glass, fabrics, wood, leather, paper, plastics and metals.

The above is only a partial list, since there are other adhesives and also a unique list of adhesives for space. Although many adhesives might be listed for a given area of application, performance and quality in most cases dictates the type and price that can be paid for the adhesive.

21.3. Why Adhesives Are Used

Adhesive bonding has many advantages to offer the construction industry. No other method of attachment is satisfactory for many applications. It would be absurd to consider nailing a ceramic tile into position or using a plywood paneling that had the wood plies stapled together. Even the piece of sandpaper in the carpenter's toolbox depends on an adhesive to hold the grit to its paper backing. When all the applications of adhesives are taken into account, adhesive bonding must be considered as the most widely used method of holding materials together.

Mechanical fasteners are, by their nature, discontinuous, and thus cause stress concentrations. Even welds, considered by many to be the best method of attaching metals, are most often fillet or plug welds and, consequently, furnish only edge attachment. The adhesively bonded joint, on the other hand, furnishes a full film of adhesive over the bonded parts, which results in a more uniform stress distribution.

Adhesive bonding in many cases offers the contractor easier installation, together with the resulting savings in construction dollars. Adhesively bonded wallboard may require less plastering and sanding. Acoustical tile for residential ceilings can often be installed faster by bonding than by stapling. Insulation batts attached to wall studs can be installed faster and with a better seal by using adhesives.

21.4. Disadvantages of Adhesive Bonding

The greatest disadvantage of adhesive bonding is uncertainty. No adequate nondestructive means of testing an adhesive bond is presently available. In fully structural applications, this may well mean that the particular component is never really tested under full service load until it is placed in the structure. Then it may be too late.

Properties of the actual glue line are somewhat difficult to determine. The modulus of elasticity of the bonding layer should ideally approach the modulus value of the substrate. This matching becomes extremely difficult, especially in metal bonding. In bonding dissimilar parts, such as wood to metal, different coefficients of expansion of the substrates may be a problem. The coefficient of expansion of the bonding layer itself may be a problem.

21.5. Where Adhesives Are Used

Adhesives are used in almost every phase of the construction industry. The use of adhesives in construction began with the finishing trades. Flooring materials, wallpaper, and roofing cements were the first volume applications. Other wall coverings, such as tile, paneling, and then ceiling panels, soon followed. Adhesives have spread into both semi-structural applications, such as the placing of gypsum wallboard, and purely structural uses, such as the glue-nailed joints in wood trusses and in the lamination of large timber girders and frames.

21.6. The Nature of Adhesion

The total adhesive force holding two materials together is the sum of two factors, namely specific adhesion and mechanical adhesion. Specific adhesion is chemical. It is the molecular attraction between two materials. The actual bond may be a chemical union, such as sulfur linkages in rubber-to-metal bonding, or it may be simply an electrical attraction between electrons of the two substances. Mechanical adhesion is the bonding force provided by interlocking action. Specific adhesion can be considered as an active force holding the materials together. It is effective under tensile, shear, and peel-type loadings; whereas mechanical adhesion is passive and not very effective until acted upon by an outside force. Mechanical adhesion is most effective under shear-type loading and contributes little to the tensile strength of a joint.

The total force holding two materials together is proportional to the bond area. It is a common misconception that roughening the surface of a joint increases the strength of the bond because it provides mechanical interlocking. Current research indicates that surface roughening increases the bond area for specific adhesion, and that the effect of mechanical interlocking is minimal in many cases. However, the roughened surface is more difficult to wet with the adhesive, and this may result in discontinuities in the adhesive film. Consequently, best results are generally obtained with surfaces that are smooth but not polished.

Most surfaces are contaminated. Even so-called clean surfaces are generally coated with a thin layer of foreign material, such as an adsorbed film of gas or a thin oxide film. Clean "dry" glass may often contain a very thin water film. These films often interfere with specific adhesion and should be removed in order to obtain the best results with an adhesive. Greasy films can be removed with a solvent wipe or other chemical wash. Xylene or toluene will remove most greasy or oil films. Acetone will also give good results, but is a little difficult to handle at the construction site since it evaporates so rapidly. For metallic surfaces such as steel and aluminum, a light sanding with fine sandpaper will remove the oxide film; and then a quick wipe with a clean rag and acetone will remove the sanding dust, and the surface is ready for bonding. For porous substrates such as wood, a fine sanding and a wipe with a clean dry rag will furnish a satisfactory substrate.

Adhesives are generally furnished to the job in liquid or paste form. Some physical or chemical change is necessary to convert the adhesive to a solid. This change may be a hydration, as in a paste of portland cement and water; it may be polymerization, as with the epoxy and polyester resins; or it may be simple evaporation of volatile materials, as in the solvent-based rubber adhesives. Other methods of curing, such as vulcanization, oxidation, gelation, and pressure reduction, are also used in various applications. Since curing methods vary so widely, some understanding of the properties of adhesives is necessary in order to insure that proper curing methods will be used on the job.

Adhesives are particular. Every adhesive will not form good bonds with every type of substrate. Materials such as wood, the unglazed backs of ceramic tiles, and paper, are porous. These materials have pores or capillaries that will tend to bleed off the vehicle from the adhesive. This can be both an asset and a disadvantage. In some cases, it may result in quick tack and fast drying. On the other hand, a porous wallpaper used with the wrong adhesive could stain badly. Metal and glass surfaces have little or no capillarity and thus do not confront this problem.

The typical adhesive joint consists of two pieces to be bonded (substrates or adherends), a layer of adhesive, and two interfaces where the adhesive comes into contact with the substrate. In this broad sense, then, a sealant in a joint fits the definition of an adhesively bonded joint. However, the sealant (the adhesive layer) is generally from ¼ to ½ inch thick, whereas in adhesive bonding the adhesive layer is quite thin. Research has been conducted on optimum glue-line thickness, and the results indicate that thin glue lines usually perform better than thick ones.

Because of the nature of specific adhesion many materials have a natural affinity toward adhesion. If two thin sheets of glass, such as the slides used with an ordinary microscope, are rubbed together, they will cling together and will require some degree of tensile force to separate. Simple evidence of the effectiveness of thin films can be demonstrated using the same two slides. If a drop of oil or soapy water (which are lubricants) is placed between the same two slides and forced into a thin layer by firmly pushing the slides together, the resulting union may have a tensile strength of as much as 60 psi. These same adhesive layers would obviously be ineffective in thick glue lines.

In any type of bonded joint, some degree of pressure is usually required to keep the parts in intimate contact until the adhesive has cured. This may vary from simple hand pressure to the large presses used in the fabricating of glued-laminated wood members. The proper amount of pressure must be used for any given application. There is no general law that indicates the amount of pressure to be applied; the required pressure must simply be sufficient to force the surfaces into intimate contact. The surfaces will never bond if the adhesive does not wet both substrates. Bonding pressure depends on the flatness of the surfaces, the type of substrate, the viscosity of the adhesive, and temperature. There are numerous examples, especially in wood joints, in which insufficient pressure has resulted in a poor bond. However, there are also examples, notably in metal bonding, in which a poor bond has been traced to excessive pressure. In order to obtain the best bond with a particular adhesive, it is best to consult the adhesive manufacturer for recommendations.

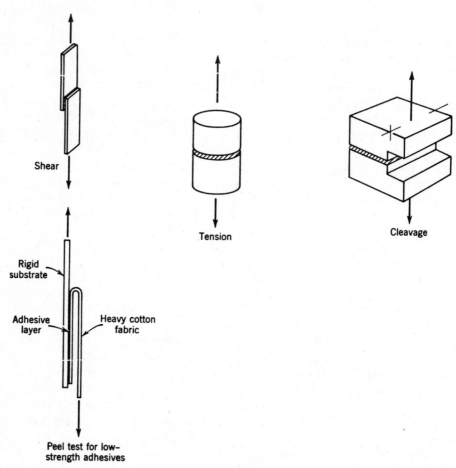

FIGURE 21.1. Adhesive test specimens.

21.7. Testing of Adhesives

Adhesive joints are tested in tension, shear, peel, and cleavage tests, and sometimes for impact strength and creep. Of these tests, the shear test is the most widely used and most widely respected. Figure 21.1 shows a typical specimen for each of the tests.

21.7.1. Shear Strength

The shear test is a measure of both specific and mechanical adhesion. However, it is conducted on a ½-inch single-lap joint that is bonded under laboratory conditions. The shear strength of metal-to-metal adhesives is determined by ASTM D-1002. Wood-to-wood specimens are tested in shear by tensile loading in a plywood construction using ASTM D-906. Shear strength of adhesives can be determined by compression loading rather than tensile loading using ASTM D-905. In most testing the values obtained from the testing must therefore be considered as maximum values. Job-site conditions will lower the values considerably. Bonded joints at the construction site involve larger bond areas and thus a risk of discontinuities in the adhesive film. Also, tolerances for the larger parts used at the job may result in an uneven glue line thick in some parts and thin in others.

The shear capacity of the adhesive is important in the fabrication and erection of sandwich panels for walls and also in the installation of bading blocks and ceramic tiles.

21.7.2. Tensile Adhesion

The tensile adhesion test may be conducted with sections of 1-inch-diameter round bars. The surfaces to be bonded are cleaned and a thin glue line is formed under moderate pressure (10 to 15 psi). After cure, the joints are tested by a tensile pull, normal to the glue line. Tensile adhesion values are a measure of specific adhesion only, and are valuable for such applications as the installation of acoustical ceiling panels. ASTM D-897 is used for wood-to-wood and metal-to-metal adhesives. ASTM D-1344 describes cross-lap specimens for determining the tensile properties of adhesives. This test was especially designed for the adhesion of glass either to itself or to other materials.

21.7.3. Cleavage Strength

Values of cleavage strength are usually less important that other test data for adhesives, and are sometimes not reported on the manufacturers' technical data sheets. The loading in the cleavage test is somewhat similar to the load inposed on the 4 × 8-foot sheets of paneling and gypsum wallboard that are installed with adhesives. However, high values of tension and shear are usually accompanied by high cleavage values; hence this test loses a little of its importance. The cleavage test is conducted by introducing a prying force at one end of a bonded specimen to split the bond apart. A suitable cleavage test is described in ASTM D-1062.

21.7.4.   *Peel Strength*

The peel strength of adhesives may be measured in several ways. For low- and medium-strength adhesives (below 1000 psi shear), the same peel-strength test used for sealants may be applicable. This test is valid because the adhesive (sealant) is rolled out into a thin glue line. For higher-strength adhesives, a small rolling drum may be substituted for the fabric substrate. Peel strength is then measured by rotating the small drum, which peels the adhesive from the substrate.

Peel strength is difficult to relate to job-site conditions. In the testing of adhesive-bonded joints, it is quite common to see a combination of peel plus tension, or of peel plus shear, as the unit fails. However, a condition of pure peel failure is seldom, if ever, seen on the job. It has been shown, nevertheless, that adhesives with low peel strength generally have low fatigue strength. Also, peel strength is related to notch sensitivity and tear strength. Consequently, discontinuities in the adhesive film are a matter of concern if an adhesive with a low peel strength is being used. The 180° test for peel strength is described in ASTM D-903 and is expressed in pounds per inch of width. The T-peel test is frequently used because of the simple configuration of the test specimen and its ease of fabrication. The T-peel test is defined in ASTM D-1876. ASTM D-2918 uses the T-peel specimen while being exposed to air, water, aqueous solutions, or relative humidity at various temperatures. The climbing drum test is used extensively in the aircraft industry, and is described in ASTM D-1781.

21.7.5.   *Creep*

Just as with sealants, the matter of creep in bonded joints has been treated far too lightly. The adhesive, once it is placed into use, is usually kept under a sustained load. The bonded parts are expected to stay in place for the life of the building. In some applications, such as floor tiles, creep is of negligible importance. However, in structural applications, such as glued-laminated members and metal bonding, creep can be the ultimate design criterion.

The creep test and its companion, the stress rupture test, can yield a great deal of information about the time-dependent properties of a bonded unit. The creep test is performed on the same type of specimen used for the shear test. For structural applications, the specimen should be loaded to 50% of its ultimate shear strength. The load is left on the specimen, and the curve of deformation versus time is plotted.

The stress rupture test is somewhat the converse of the creep test. The object of the stress rupture test is to determine the maximum load the specimen will sustain for an indefinitely long period of time. This obviously requires a series of specimens tested under different loads to a time-dependent failure. From the results of this test, the manufacturer is able to tell the architect or engineer that a certain adhesive is able to sustain a specified amount of shear loading for an indefinitely long period without failure. The stress rupture test, when used together with the creep test, which gives the amount of deformation, provides the architect with an excellent picture of the load-carrying capabilities of an adhesive.

ASTM D-2293 determines the creep properties of adhesives in shear by compression loading, and ASTM D-2294 determines the creep properties in shear by tension loading. Creep is determined by observing the displacement of five razor scratches across the centers of both sides of the lap joints with a calibrated microscope.

21.7.6. Service Life

Adhesively bonded joints must perform under various conditions of weather, temperature, humidity, and other factors. Accelerated testing can be done in weatherometers, and ASTM D-1828 tests under temperature, humidity, and light exposures. Salt spray tests are useful for evaluating adhesives to be used in coastal areas; ASTM B-117 is an appropriate test for this environment. Cyclic testing at various temperatures and humidities is covered by ASTM D-1183 and the extent of any deterioration in any test is determined by the change in strength values in shear using ASTM D-906.

There are any number of adhesives based on animal and vegetable components that are subject to attack by molds, fungi, and bacteria as well as rodents and insects. ASTM D-1286 tests the stability of packaged glues with regard to several standard organisms. ASTM D-1274 tests the bonds after exposure to various bacteria. ASTM D-1383 tests the susceptibility of films to laboratory rats, and ASTM D-1382 tests the susceptibility of films to cockroaches.

There are other test environments to which the adhesive bond might be exposed, and sample bonds can be subjected to accelerated conditions of that environment and then values in shear compared to original values. A vast number of papers have been written on exposure of bonds to various kinds of radiation (32).

21.8. Reference Standards

At least 50 different standards for adhesives are applicable to the construction industry. In addition, standards such as those of the American Institute of Timber Construction carry as much weight as federal specifications within a given trade. There are federal and military specifications, ASTM standards, ANSI standards and the standards of various groups such as the Timber Institute, the Tile Council of America, and many others. Consequently, even the experienced specifier finds it extremely difficult to have even a bare working knowledge of the requirements of all of the adhesive standards.

Fortunately, in the construction industry there are virtually no adhesive standards that cross building trades. Requirements for tile-setting adhesives are determined by one group, and adhesives for wood by a separate group. The federal and military specifications are used as guides by the specifications groups in the various trades, but rarely, if ever does the architect specify an adhesive. Rather, the architect would specify the adhesive recommended by the tile manufacturer, in order to place responsibility for the finished job on the manufacturer or the installer of the tile. The architect, in specifying the use of glued-laminated timber members, would refer to the standards of the American Institute of Timber Construction. The laminator of the structural members then assumes responsibility and uses adhesives that meet AITC standards.

The exception—the case in which the adhesive is usually specified— is in the bonding of cementitious materials. The architect specifies the bonding agent to be used in the installation of stone, quarry tile, and cement floors. The highway or bridge engineer specifies the adhesive to be used for the repair and rehabilitation of damaged concrete and other structural bonding. The adhesive most often specified is an epoxy. The specifying authority may be an architect, consulting engineer, or materials engineer of a state highway department. There are federal and military specifications that cover the epoxies. Consequently, the finished specification will usually be a reference-type specification, either based on the federal specification or else a performance specification written around the properties of a particular product that has proved itself in test installations.

The tile adhesives are often hallmarked by the Tile Council of America. The TCA maintains a research institute that assumes much of the responsibility for the quality of the adhesives. The hallmarks are related to existing standards, such as Department of Commerce CS 181 (ANSI 181). The main features of these standards are a shear test and a test for the durability of the bond. The shear test is conducted under four sets of conditions:

1. 16 hours after bonding.
2. 7 days after bonding.
3. 7 days air drying followed by 7 days of water immersion.
4. 28 days after bonding (40 psi shear).

The durability test for the quality of the adhesive bond is a cyclic test modeled after ASTM D-1037.

The lumber industry depends on specifications based mostly on research and testing done by the Forest Products Laboratory of the Department of Agriculture. Groups such as the Douglas Fir Plywood Association and the American Institute of Timber Construction base their standards on federal and military specifications developed by FPL. Ninety percent of wood gluing applications employ one of two varieties of glue—casein glue, which is used for interior work with low-humidity environments, and resorcinol glue, which is used for exterior work or high-humidity applications. Casein glues are covered by federal specification MMM-A-125. Resorcinol glues are covered by federal specification MMM-A-181 or military specification MIL-A-5534A. These wood specifications are, at present, captive specifications.

There is little likelihood that the architect will ever specify the type of adhesive used to hold plywood veneers together. However, a great deal of research is being conducted into glued-nailed trusses and glued subfloorings, and therefore architects may have to familiarize themselves with these specifications in the near future.

In the field of metal bonding, the aircraft industry has made use of the following military specifications for structural metal-to-metal bonding of structural sandwich materials:

MIL-A-5090 for adhesive, airframe, structural, and metal-to-metal applications.

Mil-A-25463 for adhesive, metal-to-metal, structural, and sandwich material applications.

These specifications appear to be quite adaptable to the needs of the construction industry. Metal bonding is beginning to show up in building construction, and the architect will in the future become responsible for specifying the proper adhesive.

22

Flooring Adhesives

Adhesives for flooring materials are high-volume, lower-cost materials. These materials are generally manufactured by the same companies that manufacture the flooring itself. The adhesives are advertised and promoted as a unit with the flooring and are generally specified by the architect. The type of adhesive varies with the type of flooring, and also with the subflooring over which it is installed.

There are nine types of flooring materials that require adhesives for their installation. Table 22.1 lists the type and their estimated 1981 sales.

In 1981 almost 25 million pounds of adhesives were used for the installation of flooring materials; 60% of the flooring adhesives were in the replacement market, and roughly 40% were used in new construction. The flooring materials use various adhesives: linoleum paste, rubber, asphalt, neoprene, nitrile, SBR, polyvinyl acetate (white glue), rubber-based mastics, resin-based mastics, neoprene contact cements, and rubber emulsions.

The substrate for a flooring materials is generally either a wood subfloor or a concrete slab. Wood subfloors are seldom used in below-grade applications, but concrete slabs may be used either below or above grade. Consequently, adhesion to wood and concrete and sensitivity to moisture are the prime considerations in the selecting of a flooring adhesive. When concrete slabs are in contact with the ground, the slabs tend to remain moist. The moisture present in the slab is alkaline in character. This alkalinity of the slab surface may have an adverse effect on adhesion.

Even though flooring materials and adhesives vary widely, a few general principles should be observed. For installations on wood subfloors above grade, there is some latitude in the choice of adhesives. Asphalt tile and parquetry wood blocks can successfully be installed with asphalt-based mastic

TABLE 22.1.   Types and Estimated Volumes of Flooring

Floor type	Millions of Square Feet
Vinyl asbestos tile	2,500
Inlaid vinyl sheet goods	1,200
Asphalt tile	1,150
Linoleum sheet	575
Wood block	350
Vinyl tile	150
Rubber tile	145
Linoleum	115
Cork tile	15

adhesives. However, asphalt-based adhesives should not be used with rubber or vinyl products. Rubber-based mastics, resin-based mastics, and white glues can be used to install both sheet goods and floor tile units. Contact cements such as the neoprenes are also very successful in the installation of all types of floor tiles, including parquet blocks. Neoprene contact cements are also the best choice for the installation of molded vinyl and rubber stair treads

FIGURE 22.1.   Installation of a sheet vinyl flooring using a mastic adhesive. (a) Spreading the mastic; *(continued)*

FIGURE 22.1 (b) Rolling the flooring; *(continued)*

and mats. Sheet flooring, such as linoleum or inlaid vinyl, is best laid with a mastic-type adhesive. Contact coments are difficult to work with when handling large sheets of flooring material. Figure 22.1 shows the installation of a sheet vinyl flooring using a mastic adhesive.

For installations on wood subfloors, the mastic adhesives are generally spread with a notched trowel. A trowel with $1/16$ to $1/8$-inch deep notches spaced on $3/8$-inch centers will serve adequately. A clean subfloor is required, but no priming of the surface is necessary. A lining felt is sometimes placed over the subfloor before installation of the flooring material. The contact adhesives may be installed by notched trowel, but are also available in a brushable consistency. In some cases, it may be necessary to apply a thin brush coat of the adhesive to seal the subfloor and prime the surface. Wood parquet blocks are now available with the contact adhesive already on the wood block. If the underlayment of hardboard or plywood is sound and true, the release paper can be peeled from the back of the blocks and the blocks can be set immediately. If the underlayment is rough, it should be brush primed, whereby much of the advantage (and expense) of this type of flooring is lost.

In residential construction much of the resilient flooring is installed on concrete subfloors, which may be slabs resting on the ground or basement

FIGURE 22.1 (c) Finishing the seams. (Courtesy of Armstrong Cork Company.)

slabs that are below grade level. In commercial and industrial construction, the majority of applications on concrete slabs are above grade level. In all applications of flooring to a concrete substrate, the major factors affecting adhesion are moisture and alkalinity. However, these problems are much more severe in the on-grade and below-grade specifications. According to Rohrer (33), one or more of the following situations may be encountered:

1. The adhesive film may be dissolved or chemically attacked by the alkaline moisture.
2. If a subfloor is wet at the time the installation is made, the adhesive will not dry or set properly. In this case an adequate bond is never obtained.
3. The dry adhesive film, even though resistant to alkaline moisture, may be stripped from the subfloor by water. The result is that the adhesive is really floating loose on a layer of water.
4. The adhesive film may be softened and stripped from the subfloor by moisture. Subsequent foot traffic will then force the adhesive through the seams or joints of the floor covering. This effect is more likely to occur with the asphalt adhesive used to install asphalt tile and vinyl asbestos tile.

Some architects have used membrane waterproofing under slabs on grade to prevent the entry of moisture into the slab. Heavy seal coats have also been used on top of the slab to provide a moisture barrier. For concrete slabs on or below grade, water-sensitive adhesives, such as polyvinyl acetate, should never be used. Research has shown that the rubber-based mastics with a water vehicle (rubber latex) will give good performance in these critical applications. For concrete slabs above grade, if the slab is permitted to dry thoroughly, more latitude is possible in the choice of adhesives. Depending on the type of floor covering being used, resin-based mastics, asphalt-based adhesives, and polyvinyl acetate may be used.

Wood block flooring (end-grain blocks) for industrial floors presents a specialized application. For large installations (over 2000 square feet) these blocks are usually set in a hot coal-tar pitch. The pitch is heated to 400 to 425°F, and spread by squeegee. This pitch has a very short "open time" (less than 1 minute), so the melted pitch must be spread rapidly. After the pitch has set, the blocks are placed and the surface is flushed over with a filler, applied by squeegee to fill all the joints in the floor. For small installations and occasional replacement of a block, a viscous asphalt emulsion can be troweled onto the subfloor. After the adhesive has dried sufficiently to become quite tacky (about 1 hour), the block may be set.

Any flooring installation is no better than the surface to which it is applied. Concrete floors should be as dry as possible. Grease and oil stains should be removed. Any loose dirt and cement dust should be removed by sweeping or vacuum-cleaning, and any protrusions on the concrete slab should be ground off to the level of the slab.

Flooring adhesives are typically low- or moderate-strength adhesives (less than 1000 psi shear). The loading on flooring adhesives is compression with little or no shear. The adhesive may be subjected to some flexing when installed over wood subfloors. Virtually all installations are interior use; consequently, thermal expansion and contraction are not too much of a problem.

23

Wall
Covering Adhesives

23.1. Ceramic and Plastic Tile

Ceramics in various forms have been used in construction since ancient times. In this country glazed, or ceramic, tile units have been used as a wall covering for almost 100 years. The tile units are made from finely ground and prepared clay which is bisque-fired to a hardened but somewhat porous state. The faces of the tile units are coated and fired again (vitrified), so that one surface has a hard, glassy finish. The backs of the tile units remain unglazed and somewhat porous.

In the early days of construction the tile-faced wall was considered in the same category as any other faced masonry wall. Consequently, the tiles were set in the same manner that a brick or stone facing was set—in a bed of mortar. Sometimes ordinary lime mortar was used, but it soon became standard practice to set the tiles in a portland cement mortar. Since the backs of the tiles were porous, some slight advantage was obtained from mechanical adhesion. Portland cement tile installations were, and still are, a quite satisfactory method of setting ceramic tile. Within the last 20 years, however, adhesives have been widely adopted for tile setting. Adhesives are faster than the mortar method and they place less dead load on the wall structure. In residential construction, adhesives are better suited for the installation of ceramic tile over gypsum wallboard.

Ceramic tiles offer many unique advantages as a wall covering and consequently the market is still growing. In 1981 approximately 300 million square feet of ceramic tile were used by the construction industry. The adhesives used for the installation of these tiles are either solvent-based rubber mastics or latex emulsion systems. The adhesives for wall tiles are

283

FIGURE 23.1.   Shipboard installation of ceramic tile using a two-component neoprene-based adhesive. (Courtesy of Crossfield Products Corp.)

not manufactured by the same companies that manufacture the tiles. However, tile manufacturers will recommend specific adhesives and architects will generally specify in accordance with these recommendations.

Wall-tile adhesives are typically low to moderate strength, low-cost adhesives. Latex emulsions such as PVA sell for about $6 to $8 per gallon. The solvent-based mastics are slightly more expensive.

Wall-tile adhesives should have a fast grab, but since the adhesive must support only the dead load of the tile, high shear strength is not required. ANSI standard 181 requires only 40 psi in shear. The adhesive should have enough resistance to creep so that the tile will not slip out of alignment. However, the shear load of the tile on the bond line is not excessive, so that both the solvent and emulsion adhesives meet this requirement readily.

There is one specification requirement for the wall tile adhesives that is quite demanding. ANSI standard 181 requires a shear test of the adhesive after 7 days of drying and 7 days of water immersion. This is an excellent specification requirement, because ceramic tile is often used for bathrooms, locker rooms, showers, and other installations in which it is exposed to high humidity conditions. The solvent-based adhesives usually meet this requirement, but few of the latex emulsions are able to pass the test.

In extremely critical water-immersion installations, such as swimming pools, a two-component neoprene-based adhesive may be specified. This type of adhesive has a higher shear strength than the other ceramic tile adhesives, as well as better water-immersion resistance. This adhesive may also be a good choice for exterior ceramic tile applications. Figure 23.1 shows a shipboard installation of ceramic tile using neoprene adhesive.

Plastic tiles have been in use in residential construction since the early 1940s. The tiles are light in weight, and easy to handle and install. However, plastic tiles are not as durable as ceramic tiles. A different set of standards covers the adhesives for plastic tile, and a different hallmark is issued by the Tile Council of America. The water immersion resistance required of plastic tile adhesives is the same as for ceramic tiles. The load-bearing capacity required of plastic tile adhesives is less than for ceramic tiles, because the light-weight tiles do not exert as much dead load on the bond line.

23.2. Wallboard and Panel Adhesives

Wallboards and paneling, in the majority of cases, are still installed by nailing to studs or furring strips. As better adhesives have developed, adhesive installations have taken over a slowly increasing segment of the market. Adhesives do offer definite advantages in many cases. Paneling may have an expensive veneered surface of wood or a decorative laminate in which nail holes are almost impossible to disguise. Gypsum wallboard, when nailed into place, requires that the nail holes be in-set and then spackled over and sanded before painting.

Adhesives are currently used for the installation of gypsum wallboard, plywood (including paneling), hardboard, vinyl-clad steel laminates, asbestos cement board, and low-density insulating fiberboard. The substrates to which these board are attached may vary, but wood is by far the most common. The adhesives, used, then must have good adhesion to wood as well as to the type of wallboard being installed. Such adhesives include mastics, contact cements, and white glue (PVA). If the building wall or furring line is fairly smooth and true, contact adhesives may be used. If the furring strips are ragged, or if a rough-textured wall such as concrete block is being covered, an adhesive such as the mastic or white glues—which have a gap filling capability—must be used.

The mastics are high-solids-content adhesives with a fast grab. They may be based on asphalt or of a reclaimed rubber resin blend. The vehicle for the rubber adhesives is a very volatile solvent, and thus good ventilation is required during installation. The rubber mastics are more expensive than the asphalt-based materials, but their lighter color offers an advantage in terms of both spotting the finished surface and operator's clean-up. With regard to strength, either type will do an adequate job. The mastics may be installed by caulking gun, knife, or trowel, but the caulking gun is the most popular method among tradespeople. Figure 23.2 shows a typical

FIGURE 23.2. Application of a drywall adhesive with air-operated caulking gun. (Courtesy of Pyles Industries, Inc.)

caulking-gun installation. Figure 23.3 shows a special long-barrel caulking gun favored by drywall applicators for ceiling and wall work without scaffolds.

The advantages of mastics include fast grab; enough body to compensate for surface irregularities; the need to apply them to only one of the mating surfaces, since they are a relatively thick glue-line application; the possibility of moving slightly for proper positioning after the panels are placed; and relative low cost.

However, the mastics may often require some bracing to hold them in position until the glue comes up to strength, and they may also require some supplemental edge nailing. The mastics are not well suited for thin-film applications, such as the installation of semirigid decorative laminates.

Contact cements, on the other hand, are intended for thin glue-line applications. They are generally viscous liquids with a neoprene base and are applied by brush or spray. Contact cements contain a high percentage of solvent for easy spreading. After the adhesive has been applied, the solvent must be allowed to evaporate before the panel is placed. The surface of the contact adhesive may be tested by lightly touching the surface with a finger. If the adhesive has lost its tacky "feel," the panel may be positioned and pressed with moderate hand pressure or a clean roller.

The advantages and disadvantages of contact cements are almost the

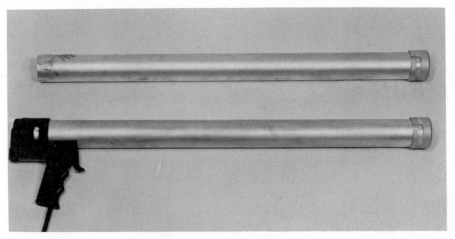

FIGURE 23.3. Long-barred caulking gun for adhesive installation on ceilings. (Courtesy of Pyles Industries, Inc.)

converse of those for mastics. Contact cements provide an instant bond with no supplemental bracing or edge nailing, as well as a good bond with a thin glue line. However, it is not possible to move the panel once it is placed, so that alignment of large sheets may be quite difficult. Because of the thin glue line, the adhesive is usually applied to both of the mating surfaces. Also, contact cements should be used only where the substrate is sound and true to line, because these glues have no appreciable thickness for filling gaps.

The white glues are emulsions of polyvinyl acetate, which is easily applied by knife, trowel, or caulking gun. These adhesives have enough strength to do an adequate job, but do not have the fast grab of the other adhesives and also are not suitable for high-humidity applications. These adhesives offer the advantages of easy application and fast clean-up. Hands, tools, and spills can be cleaned quickly with warm water. However, white glues take longer to set and, consequently, require bracing of the panels until the glue comes up to strength. White glues can be used for thick-film applications if the mating surfaces do not conform perfectly.

23.3. Other Wall Coverings

Decorative semirigid sheets of plastic laminates, such as Formica, have been used for many years as coverings for countertops and work surfaces. In recent years, however, the use of these laminates has expanded. Some of the plywood paneling available on the market has plastic laminate surface instead of wood veneer. Kitchen cabinets made from flake board or particle board coated with plastic laminates are now available. Sheets of plastic laminate approximately $\frac{1}{16}$ inch thick are now used as wall coverings. Entire kitchens can now be ordered in which the kitchen cabinets and the wall covering use the same laminate with a wood-grain finish. The same plastic laminates are finding increasing use as a wall covering in laboratories, classrooms, corridors, and other hard-usage areas.

These semirigid laminates can be installed successfully only by using adhesives. The adhesives usually used are contact adhesives based on neoprene. The laminates are supple enough to bond smoothly over the average plaster or gypsum-board wall surface. The laminates are delivered to the job in 4 × 8-foot sheets and are installed in relatively large pieces. Once the sheet is in place, the dead load stress on the bond line is negligible. There is no current reference standard for the contact adhesives used with the plastic laminates.

In the field of flexible wall coverings, wallpaper held in place by the familiar starch-based wallpaper paste remains the standard. The average homeowner would not hesitate to paint his or her own living room, but would call in a professional if the walls were to be papered. The versatility of wallpaper has greatly expanded. Papers with a washable plastic finish that are installed with wallpaper paste are currently available. Other materials that fit roughly in the wallpaper category are fabrics such as burlap and plastic-coated fabrics. Modern decorators are now even using carpeting as a wall covering for playrooms and recreation areas. These wall coverings are generally installed with a rubber-based adhesive with a fast grab. Contact cements are not recommended because the wall coverings are flexible and the pieces must be shifted slightly on the wall in order to match patterns.

The adhesive and wall covering industries have devoted considerable effort to capturing the "do-it-yourself" segment of the wall-covering market. Precoated wallpapers of various types are now available. In recent years, vinyl- and plastic-coated paper wall coverings, precoated with a contact cement, have met with success. These contact wall coverings are packed in rolls with a protective release liner. These coverings are tough and washable, and homeowners are using them for covering counter tops, furniture, and even books, as well as walls. Finishes are available in many colors, wood simulations, and imitations of brick and stone.

Compared to ordinary wallpaper, these coatings are expensive. In addition, because they bond with contact cements, positioning is critical. The piece cannot be moved once it is placed. Because of these limitations, use of this adhesive product is generally limited to small areas.

24

Roofing Cements

The roofing industry accounts for the largest volume of adhesive products used by the construction industry. The line of distinction between sealants and adhesives can be very fine, and this is especially apparent in the roofing industry. Roofing cements are mainly asphalt- or coal-tar-based materials, but with the advent of single-ply roofing, synthetic-rubber-based adhesives are also used. Some of the repair materials may be rubberized for more flexibility. The use of the roofing cements spans the entire construction industry, including both residential and commercial construction.

In residential work, the primary roofing material is usually shingles, nailed into position. However, roofing cements are used for the cementing of vent stacks, valley and chimney flashing, and eaves and gutters. The roofing cements used for residential work are usually solvent-based asbestos or fiber-filled materials applied by knife or trowel. In recent years, an adhesive rubberized liner has been offered to line leaking gutters and restore them to proper use.

The shingles used in residential work may be slate, wood, or asbestos cement, which are rigid; or asphalt shingles, which are quite flexible. The asphalt shingles are the most widely used because of their ease of installation and low cost. Asphalt shingles that contain a tab of adhesive on the lower surface of the shingle butt are currently available. This adhesive holds down the leading edge of the shingle to prevent it from being blown back and folded over in very high winds. Once the process starts, it spreads to wide areas and entire roofs have been ripped off in gale winds.

In commercial and industrial construction, hot-applied roofing cements account for the greater volume. However, in recent years cold-applied systems have been developed that are equally fast in setting. Suppliers claim

FIGURE 24.1.   Remedial job using cold-applied mastic between plies of roofing felt installed over insulation board. (Courtesy of Tremco, Inc.)

that these materials are safer to use and also eliminate any fire hazard associated with hot kettles. In many instances, the cold-applied systems have been used in remedial work where only portions of roofs are removed. Another new product in recent years is the one-sheet membrane used either for new areas or (primarily) for remedial areas, where leaky roofs are too costly to repair or remove and replace, and the sheet membranes are placed over insulation board that is anchored to the roof. The cold-applied solvent-based materials may also be used to cement cant strips, roof drains, vent stacks and flashing, while the hot-applied materials are used for cementing the bulk of the roofing material into place. For example, a building 100 × 100 feet contains 10,000 square feet of roof surface or 100 "squares" of roofing. A 5-ply built-up roof for this building contains 6 layers of adhesive, one for each layer of roofing felt and one for the aggregate surfacing. Depending on waste and layer thickness the building might use 1000 gallons of roofing cement.

Hot-applied roofing adhesives should be heated in an oil-jacketed kettle. A temperature of 400 to 450°F is sufficient to make the material fluid enough for easy mop application. Overheating is not as critical as the hot-poured highway sealants, which contain substantial amounts of rubber; nevertheless, roofing sealants should not be overheated. Temperatures above 450°F will degrade the roofing cement. Hot cements are generally

FIGURE 24.2. Hypalon sheet being installed over insulation board. This will be ballasted. (Courtesy of Tremco, Inc.)

FIGURE 24.3. Infrared sensing device used to determine wet areas to be replaced. (Courtesy of Tremco, Inc.)

FIGURE 24.4. Removal of stone and dirt prior to resaturation. (Courtesy of Tremco, Inc.)

FIGURE 24.5. Resaturating dried felt in remedial roofing application. (Courtesy of Tremco, Inc.)

applied by mopping. The roof deck is first mopped with a layer of roofing cement. The roofing felt, supplied in rolls, is placed in strips and the laps are mopped to insure a watertight proof. The final coating of roofing cement is generally thicker than the other layers in order to provide the waterproofing layer and to give a good "bite" on the aggregate surfacing.

Cold-applied adhesives have been developed for use where—because of poor access, fire hazards, or safety concern—it is impossible or very difficult to use the hot-applied roofing adhesives. These materials are very fast-setting asphalt based with high adhesive strength, and can be sprayed, brushed, or squeegeed between plies of roofing felt or fiberglass mats. Figure 24.1 shows a remedial job using cold-applied mastic over insulation board.

Another approach today is the use of single-ply membranes as roofing materials. These membranes can also be applied over old roofs. One typical method is to first make any needed repairs in the old roof. The fiberboard insulation is installed. The sheet membrane is adhered to the fiberboard with a rubber-based adhesive, and the sheeting overlaped and the edges adhered with a specially designed lap adhesive. Finally, a band of lap sealant may be laid down over the selvage edge and the adjoining sheet to ensure a watertight seam. The membrane can be a white Hypalon membrane for good light reflectance and also excellent weather resistance. Figure 24.2 shows the application of a Hypalon sheet to insulation board.

FIGURE 24.6. Embedding glass matting in a mastic to repair cracked flashing. (Courtesy of Tremco, Inc.)

Maintenance of existing roofing is another large industry. The detection of leaky roof areas has been improved on to a very high degree by the use of infrared photography and other sensors (Figure 24.3). Wet and water-saturated areas of roof show up as areas with high heat absorption; hence these areas can be located and the deteriorated areas outlined and repaired. On large roof areas it may be possible to save considerable sums of money by proper identification of areas that need more or less specific repair treatment. Deteriorated areas can be cut out and replaced, the loose aggregate and dirt vacuumed, and then the dried and cracked felts can be resaturated with very fluid asphalts that will renovate the old roof. Then a final top coat can be sprayed, after which the cleaned aggregate is replaced along with new ballast to meet the desired specification. Figure 24.4 shows vacuuming equipment for removal of loose aggregate and dirt after which the area is sprayed with resaturants as shown in Figure 24.5. It is also possible to spray coat a white Hypalon film on existing roofs where the conditions are right. Figure 24.6 shows a dried and cracked flashing being repaired by embedding glass matt with compatible mastic.

There are other materials which are used minimally in roof construction. Sealants are sometimes used as adhesives for the bedding of skylights and the like, but in volume applications no other material competes with the coal tar and asphalt cements on a price-per-performance basis.

25

Sealants
as Adhesives

The adhesively bonded joint consists of two substrates with a layer of adhesive between them. The sealed joint consists of two substrates enclosing a bead of sealant. The apparent difference between sealant and adhesive, then, is partly in the function to be served and partly in the thickness of the bond layer. There is also the concern for cost. Although many sealants can be used as adhesives, their relatively higher raw material cost negates their selection for most adhesive applications.

Many sealants are versatile enough to be used for both applications. A latex caulk based on PVA is a very low-cost material and very versatile. This material shows sufficient high shear strength to meet ANSI 181 for tile adhesives. This is a good home repair material for sticking loose tile back into place.

Polysulfide, silicone, and urethane sealants all can be made to adhere to glass, metals, and masonry, and therefore have the necessary performance characteristics as adhesives if the price is right.

Polysulfide sealants, because of their early introduction on the sealant market, found their way into many sealant and adhesive applications. Expensive paneling for elevator cages was first adhered using a polysulfide sealant. The paneling had to be braced until the sealant cured since the sealant lacked "grab," but once cured the sealant performed admirably. One of the largest adhesive markets for the polysulfides was the use of the sealant to adhere windshields into GM cars on the assembly line. This sealant eliminated the use of trim, saved on labor, and speeded up the operation. The overall cost resulted in a saving using polysulfide. The sealant was specifically formulated to give a very short work life—on the order of 3 to 5 minutes—and a complete cure in 1 hour. Pyles Industries manufactured special mixing and metering equipment to handle the three-component sealant. The sealant was used for almost 5 years on most GM cars and some Chrysler models. The desire to cut costs and simplify the equipment led to the use of a one-component urethane, which was moisture-activated with the injection of a small amount of moisture into the sealant just before application. The equipment was simpler, the compound was one-part, and the compound was cheaper. Consequently, polysulfide was replaced because of cost savings.

The replacement market for windshield sealants—which is as large as the market for original installation—was originally dominated by a two-component polysulfide sealant. However, workers all over the country had problems in mixing and application. One-component silicone with its easier gunability and no need to mix became the prime candidate for this market, in spite of its higher cost, because it was easier to use. Thus, all three types of elastomeric sealants have been used and are being used for this one area of application, the final criterion being cost, simplicity of handling, ease of application, or a combination.

Another very large market is the insulating glass market. The sealant used is an adhesive for holding the unit together, but the sealant also must have low MVT in order that the unit perform satisfactorily for a long period of time. The first materials used in the late 1940s were chemically plasticized polsulfide rubbers and the units had to be banded. When the liquid polymer polysulfides were introduced they were a natural, since they also cured to a rubber and did not need a band. The industry today uses 30 to 40 million pounds of sealant and it is shared by several sealants. The largest share is still polysulfide, which is losing ground steadily to silicone and hot melts. The prime units today are now being made using a silicone/butyl combination where the silicone is the adhesive and the secondary seal of butyl solves the requirement for very low MVT. Polysulfides are also being used for high-quality units, but now mainly in the cheaper replacement market where performance requirements are not as stringent.

The hot melts based on butyl/resin formulations are a cheaper compound. They also are used as adhesives in the insulating glass market. Because of their tendency to sag with time their use is restricted to smaller units. The hot melts when used with automated equipment make a finished unit almost instantly on the assembly line, and thus permit large volume output and a quality unit. Some urethane sealant has been tried and used in place of polysulfide, but there does not seem to be any advantage to using urethane sealant. In summary, the lucrative insulating glass market is shared by at least three qualified materials, each taking a segment of the market with cost, quality, and method of manufacture playing a major role in the final selection of materials.

Silicone sealants, because of their fast cure and their very high recovery property, are very qualified to perform as adhesives. The silicones can also adhere to most surfaces including many plastics. The sealant can be used around the house as an all-purpose sealant-adhesive for furniture repair, metal adhesive, and even bonding of patches onto children's play clothes. Silicone is used to adhere taillight assemblies.

There are many areas on automobiles where hot-melt materials based on butyl/resin, asphalt, and other adhesives based on neoprene, SBR, nitrile, and other bases, are used. Some examples are adhering insulation, trim, paneling, vinyl sheeting, and other materials with cost, performance requirements, and speed of operation all playing a vital role in the final selection of the adhesive. Unfortunately, too often cost plays the major role, with some compromise being made on quality and performance.

Elastomeric sealants have been used in setting and bedding skylights, bonding name plates to office doors, setting anchor bolts into masonry walls, forming cast-in-place gaskets, door bumpers, and other miscellaneous adhesive uses around construction sites.

The use of sealants as adhesives on the construction job are not volume applications. They are mostly odds and ends that must be taken care of on the spot. The object here is simply to point out to the construction worker or architect that a good-quality sealant can be used to stick almost any two things together in a rush situation, and there is therefore no need to hold up the job in order to get a special adhesive. The chances are very good that the sealant will do an adequate job.

26

Pressure-Sensitive Adhesive Tapes

There is hardly a home or office in the United States that could function efficiently without a roll of pressure-sensitive tape. The familiar transparent tape is used for sealing packages, tacking up notices, repairing torn books, and for hundreds of other uses. In the shipping departments of all companies, there are other tapes based on filament glass fiber reinforced for packaging larger cartons that take considerable abuse during shipping. After seeing packages come through the mail, it is evident that strong tapes are required to take the rough treatment. Various industries have requested specific tapes for a multitude of uses and environments, and many tapes today are developed to very specific requirements. The adhesives manufacturers have kept pace with this progress and now produce scores of tapes to meet the needs of contractors and industry.

Pressure-sensitive tapes serve both as sealants and adhesives depending on the application. The tapes can be used to effect a permanent seal, but also offer easy removability for temporary construction applications such as the sealing of concrete form liners.

The tapes consist of a backing material coated with a pressure-sensitive adhesive, so that the tape can be placed with only a moderate hard pressure. Baking materials range from paper and cloth to aluminum, copper, and lead foil, to films of polyethylene, vinyl, polyester, Teflon, cellophane, polypropylene, and other plastics. Backings reinforced with yarn are common, as are combinations of backings using monofilament glass, cloth, metal foils, and paper laminates. The thickness of the tape backing varies from as low as 1½ mils (0.0015 inches) up to about 15 mils (0.015 inches). The adhesives used are rubber-based adhesives, various natural and synthetic rubber/resin blends, neoprene, acrylic, silicone, and even a soluble acrylic that is used on repulpable tape.

Tapes vary in strength from about 20 to 500 pounds per inch of tape width. Tapes with reinforced backings have the highest values of tensile strength. This tensile strength, however, is the strength of the backing material and not the adhesive strength of the bonding layer. The adhesive strength varies from 10 to approximately 80 ounces per inch of width. There are a few tapes that use high-strength thermosetting adhesives, but the tapes that are used most in construction are characterized by low-strength adhesives. These tapes yield values of 10 pounds or less per inch of width when tested in shear. For permanent applications, the strength of the backing and the strength of the adhesive layer should match fairly well. For temporary construction usage, it is generally advisable to use a tape with good backing strength but low adhesive strength to provide easy removability. In the repainting of an older building, the painter might use masking tape to mark the borders of different colors. The tape should have enough backing strength to stand abuse, so that it can be put under some tension and placed in a straight line. Obviously the tape must be removable. If the adhesive strength is too high, the tape may well pull loose some of the previous coat of paint when the tape is removed. The painter then has a time-consuming patch and touch-up job to do. Also, the tape might tear and be difficult to remove.

Tapes used for permanent applications should have good aging and weathering characteristics. Tapes for both permanent and temporary construction use usually require a high degree of resistance to water. Some tapes, especially those used in duct work, should also have good abrasion resistance.

Six tables are shown listing approximately 40 tapes grouped in several categories. The data are selected from listings of over 150 types of tapes from three major manufacturers of industrial tapes. These major manufacturers are 3M, Permacel, and Tuck Industries. There are many other smaller manufacturers who also make selective tapes to industry. Table 26.1 lists representative types of paper-backed tapes. Table 26.2 lists various glass-reinforced tapes including monofilament glass. Table 26.3 lists various cloth-backed tapes, and Table 26.4 lists representative plastic-backed tapes—of which there are a wide variety. Table 26.5 lists various metal foil tapes covering lead, copper, and aluminum. The numbers and letters used for references in all of the tables are defined in Table 26.6.

There are a number of other tapes in addition to these, including foam sealing and mounting tapes for insulating and sound deadening, and weather-resistant mounting tapes for joining two surfaces. The tapes given in the tables are representative of the more popular tapes, but do not cover specific tapes made for unique areas of application. There are silicone rubber tapes for weather-resistant high-voltage cable, a number of Teflon tapes,

TABLE 26.1. Adhesive-Backed, Pressure-Sensitive, Paper-Sealing Tapes

Description	Adhesive	Tensile Strength (lbs/inch)	Spec.	Total Thickness (mil)	Adhesion (ounces/inch)	Temp. Range	Remarks
Impregnated paper	(1)	20	(a)	7	20	A	used in making white goods, no stain
Impregnated paper	(2)	20	(a)	6.5	20	B	solvent-resistant
Kraft paper	(1)	22	(a)	6.5	20	A	paint masking tape
Impregnated paper	(1)	20	(a)	6.5	20	C	super high temp. for aircraft
Kraft paper impregnated	(1)	40	(b)	9.5	20	B	strongest paper tape for industrial applications
Kraft paper impregnated	(1)	50	(c)	9.5	20	A	paper packaging
Kraft paper	(1)	50	—	10.5	22	A	double-faced for mounting glass, metal, or polyethylene; splicing
Repulpable tissue	(7)	—	—	6.5	22	A	static and low-pressure splicing, repulpable

TABLE 26.2. Adhesive-Backed, Pressure-Sensitive, Monofilament Glass Reinforced Tapes

Description	Adhesive	Tensile Strength (lbs/inch)	Spec.	Total Thickness (mil)	Adhesion (ounces/inch)	Temp. Range	Remarks
Paper base	(1)	310	(d)	12.5	48	A	for meat packaging, meets USDA specs
Polyester base	(2)	300	(e)	10	40	A	shock absorbing for metal surfaces and conduit bundling
Polyester	(1)	380	(f)	8	40	A	high-tensile tape for maximum security for heavy loads
Polyester	(1)	340	(g)	7.5	40	A	general-purpose, USDA approved
Polypropylene (nonreinforced)	(3)	180	—	4.7	24	A	for medium-load palletizing
Glass cloth	(4)	170	(h)	7.5	25	C	high-temp. electrical, UL approved

TABLE 26.3. Adhesive-Backed, Pressure-Sensitive, Cloth Tapes

Description	Adhesive	Tensile Strength (lbs/inch)	Spec.	Total Thickness (mil)	Adhesion (ounces/inch)	Temp. Range	Remarks
Plain cloth	(1)	40	—	12	40	A	multipurpose
Plain cloth	(5)	55	—	12.5	35	A	high-strength, weatherproof
Vinyl coating	(1)	50	(i)	14	40	A	extra-strength, waterproof
Polyethylene coated	(5)	50	(i)	11	45	A	water- and weatherproof, bundling and carton sealing
Polyethylene coated	(1)	35	—	11	45	A	duct work sealing, waterproof
Plain cloth	(2)	100	(j)	17	75	A	double-faced, flame-retardant for aviation industry

TABLE 26.4. Adhesive-Backed, Pressure-Sensitive, Vinyl and Plastic Tapes

Description	Adhesive	Tensile Strength (lbs/inch)	Spec.	Total Thickness (mil)	Adhesion (ounces/inch)	Temp. Range	Remarks
Vinyl film	(2)	20	(s)	7	20	A	electrical tape, dielectric strength 9,000 volts
Vinyl film	(2)	20	(j)	7	20	A	all-weather electrical tape, dielectric strength 9,000 volts wet
Vinyl film	(2)	18	(k)	11	20	A	premium electrical tape, 8 color codes, OSHA approved
Vinyl film	(2)	32	—	11	20	A	aisle masking tape
Vinyl film	(2)	28	—	10	20	D	pipe covering below ground, IAPMO listed
PVC film	(2)	19	—	8.5	60	A	double-faced for carpeting and vinyl tile
Polyester	(6)	20	—	16	20	B	for splicing silicone papers and interliners
Teflon	(6)	20	(1)	3.5	20	B	for release wraps on rollers in paper mills
Teflon	none	7	—	3.5	—	D	ribbon dope thread sealant, friction reducing
Polyethylene	(1)	45	(m)	6.5	20	A	friction reducing movement of any material
Cellophane	—	25	—	3	30	A	transparent, used over nameplate
Polypropylene	(3)	20	(r)	2.2	20	A	brilliantly clear for film splicing

TABLE 26.5. Adhesive-Backed, Pressure-Sensitive, Metal Foil Tapes

Description	Adhesive	Tensile Strength (lbs/inch)	Spec.	Total Thickness (mil)	Adhesion (ounces/inch)	Temp. Range	Remarks
Aluminum foil	(7)	29	(n)	6.5	50	B	ultimate in moisture vapor protection
Lead foil	(2)	20	(o)	8.5	35	B	very low MVT, used in electroplating, chemical radiation barrier
Lead foil	(1)	17	—	5	40	B	for use in simulating leaded glass windows, also X-ray shielding
Aluminum foil	(1)	50	—	8	65	B	protect plastic wiring in refrigerators, heavy duty
Aluminum foil	(6)	75	(p)	6	22	C	high-temp. performance, highly reflective to radiant heat
Copper foil	(1)	32	—	3	45	B	conducting tape, low MVT

TABLE 26.6. Explanations for Tables 26.1 to 26.5

Adhesives	Specifications	Temperature Ranges
(1) synthetic rubber/resin	(a) PPT-42C-I	A 32 to 180°F
(2) natural rubber/resin	(b) PPT-42C-II	B can take 2 hours at 300°F
(3) acrylic	(c) PPT-76C	C can take 5 hours at 350°F
(4) rubber thermosetting	(d) PPP-T-97B Type IIA	D 35 to 225°F
(5) synthetic rubber	(e) PPP-T-97E Type I	
(6) silicone	(f) PPT-97D Type IIB	
(7) water-soluble acrylic	(g) PPT-97E Type IIB	
	(h) MIL-I-5126F-GFT	
	(i) PPP-T-60D Type IV Class I	
	(j) GSR-Part 25	
	(k) PPP-T-66dI	
	(l) MIL-I-23594 I	
	(m) FDA-21CFR	
	(n) L-T-80B	
	(o) HH-T-0029	
	(p) MIL-T-83284-USAF	
	(q) PPP-T-70	
	(r) ASTM-D-2301 I	

motion picture perforated film splicing tapes, innumerable decorative tapes in a wide variety of colors, tapes for the graphic arts, many double-faced tapes, non-yellowing clear plastic tapes, many repulpable tapes for the paper-making industry, high-temperature tapes for the paper-making industry, high-temperature tapes, flame and radiation resistant tapes, and many others. The tape manufacturers can manufacture tapes that can meet any number of unusual requirements as long as the volume is right.

26.1. Testing and Reference Standards

Tapes for construction and industrial use are a relatively new product in terms of usage volume. Consequently these products are being intensively tested by the manufacturers before marketing. The usual tests conducted on adhesive tapes are as follows:

1. Tensile strength of the backing material in pounds per inch of width.
2. Shear strength of the adhesive in pounds per inch of width.
3. Peel strength of the adhesive in ounces per inch of width; usually tested to steel, but other surfaces can be specified.
4. Aging and weathering, usually covered in a specification.
5. Creep of the adhesive under constant load.
6. Percent elongation on some conformable plastic tapes.
7. Light transmission of various wavelengths for lithographic tapes.
8. Dielectric strength, when required.
9. Flame and weather resistance, when required.

Manufacturers have developed their own performance requirements for the tapes. Reference standards have been written for many types of tapes. There are enumerable federal specifications, FDA specifications, ASTM standards, and industrial standards. Almost all tapes supplied by manufacturers are described with respect to various performance parameters along with end-use application, so that anyone looking for a specific type of tape could undoubtedly find a type meeting most of the requirements.

The manufacturers list their various tapes by their own nomenclature so that architects and contractors can specify and buy tapes by brand name to minimize confusion. This buying has channeled the tape market to a relatively few large manufacturers who produce a wide range of products.

26.2. Applications of Adhesive Tapes

Adhesive tapes are used for both permanent and temporary construction. Applications of the tapes are so diverse that it is impossible to list them all, but enough applications are listed in Tables 26.1 to 26.5 to show what the tapes can do.

One permanent application of tapes is in the sealing of lap joints in metal roof decking. The tape for this application may be a metal foil tape to match the properties of the deck being sealed. The metal roof deck often serves as a base for a hot-applied built-up roof, so the tape adhesive must therefore have high heat resistance.

Another permanent application of adhesive tape is the coating of pipes to be laid underground. The tape is easily applied and provides permanent corrosion resistance for the pipe. Since the pipe is to be buried in soil that may be wet, both the tape backing and the adhesive must have good water-immersion resistance.

Tapes are also used to hold preformed insulation in place around piping. This application is often found in residential construction. The pipes are coated with insulation to prevent condensation. If the pipes are to remain exposed, as in typical basement, the tape backing should be paintable so that the basement will have a finished appearance.

Heating and air conditioning ducts are quite often joined with pressure-sensitive tapes. The weight of the duct work is supported by the building frame, but the tape serves as both sealant and adhesive to join the sections of duct work.

Residential builders sometimes use 2-inch wide tapes to provide draft-free floors. These tapes are used to seal the floor-to-wall joint and then are covered by the base molding.

Some tapes have become quite commonplace on the construction site. It is probably safe to say that no electrician would walk onto a job without a roll of adhesive tape in the tool kit.

Temporary uses of tapes in construction probably outnumber the permanent installations. The "closing-in" of a building during cold weather is often accomplished by large polyethylene sheets. There sheets are very conveniently jointed by 2-inch-wide strips of polyethylene tape.

Often additions to existing buildings must be made with a minimum of interference to the operation of the existing building. In this case, temporary partitions must be erected to keep construction dust out of the existing building. The most convenient way to seal these partitions is by using pressure-sensitive tapes.

The graphic arts use highly specialized tapes. Brilliantly clear tape is needed for film splicing. Photographers need a black opaque matte-surface tape for masking out areas to be photographed. There is a red-dyed polyester tape that is opaque to all colors but red for "see-through" in positioning on a light table. The drafters need an easy-release adhesive tape.

Double-faced tapes are also made for a multitude of purposes. The paper industry requires several types of repulpable tape for splicing paper during manufacture. These tapes come in widths up to 18 inches. The double-faced tapes are needed for adhering various surfaces together, including carpeting and carpet squares.

Foam sealing and foam mounting tapes are used for insulating, sound deadening, vibration dampening, and mounting two surfaces together.

There is even a 24-karat-gold-coated polyimide tape used to protect spacecraft parts and components from intense radiant heat. This has also been used in decorative trim in combination with gold leaf.

26.3. The Future of Adhesive Tapes

The use of adhesive tapes in construction has grown almost four times as fast as the overall construction industry. Contractors and industry have realized the convenience of tapes and are using them more and more. Temporary uses of tapes should continue to grow rapidly. However, tapes for permanent applications will probably show slower growth. Any component that remains as a permanent part of the structure will be subject to regulation by specifications and local building codes. This regulation will lead to standardization of manufacture and testing for certain types of tapes, which will somewhat slow down the growth. In the overall picture, the tapes should continue to grow at least twice as fast as the construction industry for the next 10 years.

27

Adhesives as Binders for Composite Materials

Composite materials are those that combine two or more dissimilar materials so that the system acts as a unit. There are dozens of examples of composite materials actively being used by the construction industry. Concrete consists of sand and stone glued together so that it acts as a unit. Particle board and hardboard are made from small pieces of wood held together by an adhesive. Composite beams are formed by casting a concrete on top of a steel beam; in this case, the two elements are held together by a shear connector so that they act as a unit. Structural sandwich panels are also composites.

Composite materials have many advantages, and the use of these materials is growing. Every material known has its strong and weak points, and composites often combine materials in such a way that the best use is made of each element. Concrete is good in compression, but weak in tension. Consequently, the composite beam places the concrete slab into compressive loading, and the steel beam section takes the tension. Composites often make use of very-low-cost raw materials and combine them into high-quality finished products. Particle board uses wood chips that are actually mill waste from the lumber industry. These chips act as a fibrous reinforcement for low-cost-phenolic or urea-resin adhesives, and are formed into sheets for countertops, kitchen cabinets, and other construction uses in which sheet materials are required. These hardboards can be sawed, shaped, and nailed just like lumber. Approximately 750 million pounds of adhesives were used by this industry in 1981. This is a captive use of adhesive, however. The adhesive used is the choice of the manufacturer. The architect or engineer never specifies the adhesive to be used in hardboard.

Thermal insulation might also be considered a composite material. Adhesives are used to secure paper backings and vapor barriers for batt-type insulation. Other types of insulation also use an adhesive to hold the fibers together. This is another high-volume use of adhesives in construction (250 million pounds in 1981); but this also is a captive use. The architect or engineer checks the heat-loss rating of the insulation and specifies the type and amount of insulation to be used, but never specifies the type of adhesive in making the insulation.

27.1.   Polymer Concrete

Concrete is an artificial conglomerate rock. It consists of sand and stone glued together. If the aggregates are glued together with a paste of portland cement and water, the result is portland cement concrete. If the aggregates are glued together with asphalt, the end product is bituminous concrete asphalt or "blacktop." If the aggregates are held together by a polymeric binder, such as an epoxy, a polyester, or an elastomer, the result is polymer concrete. Polymer concretes are just beginning to receive widespread use in construction. Approximately 7 million pounds of epoxy resin were used in 1981. The greater portion of this epoxy was used as an adhesive for concrete, but a significant percentage was used as the binder material in epoxy concrete. Other resins such as the polyesters are also being used (34).

Uses of polymer concrete have thus far been small-scale operations, mostly repair work in nonbuilding construction such as bridges. Epoxy resins can be combined with sand and used in patching material for concrete. Traditionally, concrete patches have been difficult to apply successfully. Usual practice has been to cut back the concrete area to a depth of 2 inches. Sides of the patch were cut vertically because the patches could not be feather-edged. With epoxy concrete, patching is somewhat easier. The surface must be cut back to sound concrete, because any patch is only as good as the substrate to which it is bonded. However, the patch does not have to be any specified depth, and feather-edging of patches is feasible with epoxy mortars. When epoxy mortars were first used as patching materials, frequent failures were noted in the concrete below the bond line. Many of these failures were attributed to the difference in strength and the coefficient of expansion between the portland cement concrete and the epoxy mortar. Subsequent research has developed more flexible epoxy systems that have given excellent results.

Polymer concretes are currently used to surface bridge decks, for audible warning strips ("rumble strips") in high-hazard traffic areas, and for providing extra skid resistance in high-density intersections. Parking garages with steep ramps can also use these materials to advantage. The most commonly used polymer concrete surfacing is based on epoxy resin. The epoxies used are usually given added flexibility by the addition of either a coal tar, a polysulfide, or a polyamide resin. Usual practice is to use a truck-mounted sprayer to cover one traffic lane width at a time with the resin. Graded aggregate is then spread over the surface to provide skid resistance. After the resin has cured, the excess aggregate may be swept up and reused. This type of application provides a good wearing surface with very little addition of dead load to the structure as compared with other surfacing materials. The George Washington Bridge in New York City was surfaced with an epoxy concrete wearing surface.

Polymer concretes offer certain advantages so that their use will spread into other areas of construction. The properties vary, of course, with the type of binder being used. Many different polymers are currently being investigated, including epoxies, polyesters, polysulfides, polyvinyl acetate, and a series of urethanes. Polyester and epoxy concrete give compressive strengths of 12,000 to 20,000 psi as compared to 3000 to 5000 psi for portland cement concrete. Polymer concretes also offer a fast cure—72 to 96 hours versus 28 days for normal concrete. Disadvantages of polymer concretes are the high heat generated during cure and the high cost. Also, a great deal of research must be done on creep and long-term weathering properties before architects and engineers will specify these materials for beams, columns, and slabs. The more resilient binders such as polysulfides and urethanes give low-strength concretes capable of very high deformations without failure.

Busch (24) describes various types of polymer concrete systems with specific emphasis on elastomeric concrete, which has been used in a number of installations throughout the world in transition dams and wide expansion joints. Elastomeric concrete contains up to 80% of aggregate dispersed in the elastomeric sealant.

Polyester concrete has gone high-fashion. Sinks and countertops, vanities, stair treads, tabletops, bathtubs, bar tops, and a host of other items are now fabricated of polyester resin and colored aggregate. The range of colors is virtually unlimited and the material can be custom-molded into almost any shape.

Polymer concretes are also used for beautiful terrazzo floors, and in some areas a thin-set terrazzo is adhered to the concrete floor with an epoxy adhesive.

27.2. Composite Beams

The most commonly used type of composite beam consists of a concrete slab resting on the top flange of a steel beam. Some type of shear connector is used to prevent slip between the beam and slab and thus insure that the composite beam acts as a unit. The shear connectors most frequently used are steel studs welded to the top flange of the beam. In recent years, however, considerable research has been directed towards the use of adhesively bonded composite beams. Three general approaches have been used:

1. The slab is precast and then glued to the steel beam with an epoxy adhesive.
2. The top flange of the steel beam is coated with epoxy resin and the concrete slab is cast on top of the uncured epoxy. The system then cures together.
3. The top flange of the beam is coated with an epoxy or polyester adhesive. Graded aggregate, up to ¼-inch size, is then embedded in the resin. After the adhesive has cured, the concrete slab is cast and the exposed aggregate provides the shear resistance. This approach is known as the bonded aggregate beam method.

At the present time all of these systems are experimental. However, it is probable that one or more of these systems will become economically feasible in the near future. For example, the bonded aggregate beam could be shop-fabricated and shipped to the job site with a protective liner around the top flange. One of the current problems with composite construction is that once the studs are in place, it is difficult for steelworkers to walk on top of the beams. With bonded aggregate beams, steelworkers would simply walk on top of the protective liner until the slab was ready to be cast. The liner would then be removed and the slab concrete placed.

27.3.   The Future of Composite Materials

There are actually very few pure materials used in construction. In the broadest sense of the word most construction materials are composites, which may be either natural or man-made. Wood is really a composite material that consists of a group of roughly parallel cells, held together by an adhesive. Concrete is a man-made composite. Metalurgists are currently investigating high-strength composites formed by embedding whisker fibers of one metal in a matrix of another metal. Decorative laminates such as Formica are actually composite materials. Gypsum wallboard, consisting of a gypsum core and paper surfacing, is also a composite material.

The one property that all these composites have in common is that the elements of the composite are held together by some type of adhesive. As the supply of high-grade ores and other natural materials diminishes, it is inevitable that more and more composite materials will be used in construction. The use of adhesives as an integral part of these composites can be expected to grow accordingly.

28

Adhesives for
Concrete, Cement, and Plaster

Adhesives for concrete, cement, and plaster cover a wide range of construction problems. Cured concrete may be bonded to cured concrete, as in the installation of precast traffic buttons to a highway surface. Steel bridge railings may be glued to the concrete surface of a bridge sidewalk. In the case of spalled or deteriorated concrete, the engineer may wish to rebuild the structure to its former line and grade. Spalled or damaged plaster may also be recoated to furnish a new surface. It is often desirable to use a separate cement mortar floor topping over a structural floor. This separate floor topping can be bonded successfully to existing floors of either wood or concrete. These are all different problems and consequently there is no "cure-all" answer. However, all the problems can be grouped and the possible solutions listed, so that the best possible match can be made:

1. Bonding cured concrete to cured concrete.
2. Bonding cured concrete to other materials.
3. Bonding new concrete to cured concrete.
4. Bonding new concrete or cement mortar to other materials.
5. Bonding new plaster to old plaster.

In all of these problems something must be glued to something else, and this requires the use of a bonding agent. In the bonding of cured concrete to cured concrete, the adhesive is applied to the mating surface. In the case of new concrete over old, the bonding agent may be incorporated into the new concrete, or it may be applied separately to the old concrete before the new concrete is placed. The bonding of cementitious materials can be done with rubber, acrylic, or vinyl emulsions, or with epoxies. Epoxies are

313

FIGURE 28.1. (a) Epoxy bonding agent used for installing a stone surface over a concrete slab. In this installation the epoxy is used to bond the mortar bed to the structural concrete slab. The epoxy can be seen as the dark area adjacent to the fresh mortar. (b) A finished section of the installation shown in (a). In this installation the joints were filled with a mortar consisting of fine sand combined with a very flexible epoxy.

the highest priced of the group, but have taken over the market for exterior applications where high strength is required. For indoor use, where low to moderate strength is required, any of the other three adhesives may be applied.

The straight epoxy resin adhesives are very strong. They have high specific adhesion to metals, glass, ceramics, and masonry, because of their epoxy, hydroxyl, amine, and other polar groups. When the resin is properly cured, the cohesive strength is so high that failure almost always occurs in the adherend or the interface. The epoxies cure without releasing any byproduct, and have very low shrinkage. Consequently they lend themselves to assembly line application. They are also 100% solids so that no pressure is needed during application. The resins have excellent resistance to moisture, solvents, and most chemicals, and are used in potting applications for electrical components because of their excellent electrical resistance.

Their one chief problem is that they cure quite brittle, but this can be solved by using flexibilizers. Polysulfides, mercaptan terminated polymers, polyamides, and even coal tars can be used to introduce flexibility and toughness to the epoxies. The epoxy resins are furnished to the job site as two-component materials, which must be mixed before use. The typical epoxy adhesive has a shear strength of 4000 to 5000 psi and a tensile adhesive strength of 2500 to 3000 psi. The epoxy can be applied by brush, spray, or even troweled on, and has an open time of 15 minutes to 4 hours depending on formulation and mass. In mass, the exotherm can cause the reaction to get very hot and cure in minutes. In thin films, there is no chance for the heat to get trapped, and consequently the cure time is slow. Optimum cure is reached in 1 day with an exothermic reaction and in 4 to 7 days with thin-film applications. All that is required is enough pressure to keep the parts in contact. Figure 28.1 shows the use of an epoxy resin to install a stone-wearing surface on a pedestrian bridge.

The rubber, acrylic, and vinyl emulsions are one-component adhesives. They have a tensile and shear strength of 100 to 200 psi. These adhesives do not weather as well as the epoxies in exposed locations. Available in paste or brush consistency, for indoor applications these adhesives offer the advantage of no mixing, fast clean-up, and economy. They can be applied faster and at half the cost of the epoxies.

28.1. Bonding Cured Concrete to Cured Concrete

There are many applications of this type of bonding. Precast traffic buttons can be glued to a highway. Precast sections of curb can be bonded into place on city streets. Bearing blocks and machine bases can be bonded to industrial floors. Structural repairs to concrete can be made by bonding. The bonding agent used for this application is an epoxy resin. Some polyesters have been used, but these are not as widely accepted as the epoxies. Epoxy resins are available as brush-consistency adhesives for surfaces that mate well, or as a gel for nonmating surfaces. The surfaces that are to be bonded must be clean and sound. Loose aggregate, dust, laitance, and curing compounds must be removed from the surfaces. The epoxy resins are furnished in premeasured two-component kits, which must be mixed immediately before using. Tests have shown that the bond between concrete members is almost always higher than the strength of the concrete.

28.2. Bonding Cured Concrete to Other Materials

Handrails for concrete stairways can be bonded into place. Steel handrails can be glued to a bridge sidewalk. Composite beams may be formed by bonding steel beams to precast slabs. Concrete beams can be strengthened by bonding steel plates to them (36–39). Precast slabs can be glued into place on steel purlins to form a concrete roof deck. These applications also use the epoxy resin. The application is similar to bonding pieces of concrete together. When bonding to steel, the bonding surface of the steel should be cleaned down to bare metal, preferably by sandblasting or power brushing with a wire brush.

28.3.   *Bonding New Concrete to Cured Concrete*

Patching and rehabilitation of disintegrated concrete has always been a troublesome problem for the maintenance engineer. If small volumes of patching are required, an epoxy mortar is probably the best solution. However, if larger quantities of concrete are involved, a bonding agent and portland cement concrete will probably prove to be more economical. If new concrete is to be bonded to cured concrete, the existing concrete must be clean and sound. If the old concrete shows any signs of disintegration, the surface should be cleaned with chipping hammers and sandblasted down to sound concrete. The epoxy adhesive can then be applied to the surface by brush or spray and the new concrete can be placed.

In the placement of industrial floors with large floor areas, a separate concrete topping is sometimes placed on top of the structural floor slab. These separate toppings were, for years, quite troublesome until the use of bonding agents became common. For such indoor applications, the rubber latex, acrylic, or vinyl emulsion would be suitable. For heavy use areas, it is best to use epoxy resins. The emulsions may be applied directly by brush or spray, or they may be mixed with a thin mortar and brushed into place. The epoxies may also be used for this application, but their high strength are not required. Where thin set terazzo is applied to a concrete floor, the epoxy resin must be used since the grinding discs might shear off the topping if it were not adhered with epoxy resins.

28.4.   *Bonding New Concrete to Other Materials*

New concrete may be bonded to other construction materials such as steel, stone, wood, and brick. The epoxy is probably the safest choice for all these bonding jobs, in spite of its higher cost. The epoxy provides better strength, moisture and salt resistance, and resistance to the freeze–thaw cycle than the other bonding agents can provide.

Some increase in the bonding capabilities of new concrete can be obtained by using a bonding agent as an adhesive to the new concrete. Polyvinyl acetate has been used successfully for this type of bonding (35). A thin concrete topping (⅝ inch) with a PVA admix was placed over an existing wood block floor in the Testing Laboratory at the University of Cincinnati. This topping was placed in 1954 and is still giving satisfactory service.

28.5. Bonding New Plaster to Old Plaster

This is an indoor operation that requires a good bond but demands little with respect to shear strength. The vinyl emulsions based on PVA adhesives have a shear strength that is adequate for plaster bonding. If the plaster is to be exposed to high-humidity conditions, vinyl agents are not suitable. An epoxy adhesive would function better in a high-humidity environment.

Appendix 1

ASTM Reference Standard C-920 for Elastomeric Joint Sealants*

1. Scope

1.1 This specification covers the properties of a cured single- or multi-component cold-applied elastomeric joint sealant for sealing, caulking, or glazing operations on buildings, plazas, and decks for vehicular or pedestrian use, and types of construction other than highway and airfield pavements and bridges.

1.2 A sealant meeting the requirements of this specification shall be designated by the manufacturer to be one or more of the types, classes, grades, and uses defined in Section 4.

2. Applicable Documents

2.1 *ASTM Standards*:

C 510 Test for Staining and Color Change of Single- or Multicomponent Joint Sealants

C 603 Test for Extrusion Rate and Application Life of Elastomeric Sealants

C 639 Test for Rheological (Flow) Properties of Elastomeric Sealants

C 661 Test for Indentation Hardness of Elastomeric-Type Sealants by Means of a Durometer

---

*Reprinted with permission from the *Annual Book of ASTM Standards,* Part 18, American Society for Testing and Materials, Philadelphia, 1980.

C 679 Test for Tack-Free Time of Elastomeric-Type Joint Sealants

C 719 Test for Adhesion and Cohesion of Elastomeric Joint Sealants Under Cyclic Movement

C 792 Test for Effects of Heat Aging on Weight Loss, Cracking, and Chalking of Elastomeric Sealants

C 793 Tests for Effects of Accelerated Weathering on Elastomeric Joint Sealants

C 794 Test for Adhesion-in-Peel of Elastomeric Joint Sealants

3. Significance and Use

3.1 This specification covers several classifications of sealants as described in Section 4 for various applications. It should be recognized by the purchaser or design professional that not all sealants meeting this specification are suitable for all applications and all substrates. It is essential, therefore, that the applicable type, grade, class, and use be specified so that the proper classification of sealant is provided for the intended use. Test methods relate to special standard specimen substrates of mortar, glass, and aluminum. If tests are required using substrates in addition to or other than the standard, they should be so specified for testing.

4. Classification of Sealants

4.1 A sealant qualifying under this specification shall be classified as to type, grade, class, and use as follows:

4.1.1 *Type S*—A single-component sealant.

4.1.2 *Type M*—A multicomponent sealant.

4.1.3 *Grade P*—A pourable or selfleveling sealant that has sufficient flow to form a smooth, level surface when applied in a horizontal joint at 4.4°C (40°F).

4.1.4 *Grade NS*—A nonsag or gunnable sealant that permits application in joints on vertical surfaces without sagging or slumping when applied at temperatures between 4.4 and 50°C (40 and 122°F).

4.1.5 *Class 25*—A sealant that when tested for adhesion and cohesion under cyclic movement (7.8) shall withstand an increase and decrease of at least 25% of the joint width as measured at the time of application, and, in addition, meet all the requirements of this specification.

4.1.6 *Class 12½*—A sealant that when tested for adhesion and cohesion under cyclic movement (7.8) shall withstand an increase and decrease of at least 12½% of the joint width as measured at the time of application, and, in addition, meet all the requirements of this specification.

4.1.7 *Use T*—A sealant designed for use in joints in pedestrian and vehicular traffic areas such as walkways, plazas, decks and parking garages.

4.1.8 *Use NT*—A sealant designed for use in joints in nontraffic areas.

4.1.9 *Use M*—A sealant that meets the requirements of this specification when tested on mortar specimens in accordance with 8.9 and 8.10.

4.1.10 *Use G*—A sealant that meets the requirements of this specification when tested on glass specimens in accordance with 8.9, 8.10, and 8.11.

4.1.11 *Use A*—A sealant that meets this specification when tested on aluminum specimens in accordance with 8.9 and 8.10.

4.1.12 *Use O*—A sealant that meets this specification when tested on substrates other than the standard substrates in accordance with 8.9 and 8.10.

5.   *Materials and Manufacture*

5.1 A single component chemically curing sealant shall be a homogenous mixture of a consistency suitable for immediate application by hand or pressure caulking gun or by hand tool. The sealant when completely cured shall form an elastomeric solid capable of maintaining a seal against gas, liquid, and solid.

5.2 A multicomponent chemically curing compound shall be furnished in two or more components, with the base component having suitable reinforcing agents, liquid or paste curing agents, and suitable extenders for on-the-job mixing. The resulting mixture shall be homogeneous and of a consistency suitable for immediate application by hand or pressure caulking gun, or by hand tool. The sealant when completely cured shall form an elastomeric solid capable of maintaining a seal against gas, liquid, and solid.

6.   *General Requirements*

6.1 *Stability*.

6.1.1 A single-component sealant, when stored in the original unopened container at temperatures of not more than 27°C (80°F) shall be capable of meeting the requirements for extrudability (7.2.1) for at least 6 months after date of delivery.

6.1.2 A multicomponent sealant, when stored in the original unopened container at temperatures of not more than 27°C (80°F) shall be capable of meeting the requirements for extrudability (7.2.2) for at least 12 months after date of delivery.

6.2 *Color*—The color of the sealant, after curing 14 days in a laboratory controlled at 23 ± 2°C (73.4 ± 3.6° F) and 50 ± 5% relative humidity, shall be that color which has been agreed upon between the purchaser and the supplier.

6.3 The sealant shall be intended for use only on clean, dry surfaces. Where a primer is recommended by a manufacturer for a specific surface, all tests on that surface shall include the primer.

7. *Physical Requirements*

7.1 *Rheological Properties*:

7.1.1 Grade P (pourable or selfleveling) sealants shall have flow characteristics such that when tested in accordance with Method C 639 it shall exhibit a smooth, level surface. (Refer to Types I and III in the test.)

7.1.2 Grade NS (nonsag) or gunnable sealant shall have flow characteristics such that when tested in accordance with Method C 639 it does not sag more than 4.8 mm ($\frac{3}{16}$ in.) in vertical displacement. Also the sealant shall show no deformation in horizontal displacement. (Refer to Types II and IV in the Test.)

7.2 *Extrusion Rate*:

7.2.1 Type S (single component), Grade P (pourable or self leveling) sealant shall have an extrusion time of not more than 20 s when tested in accordance with Method C 603.

7.2.2 Type S (single component), Grade NS (nonsag or gunnable sealant) shall have an extrusion rate of not more than 45 s when tested in accordance with Method C 603.

7.3 *Application Life*:

7.3.1 Type M (multicomponent), Grade P (pourable or self leveling) sealant when tested in accordance with Method C 603 shall be considered to meet the requirement for 3 h application life if the time to empty the cartridge does not exceed 20 s.

7.3.2 Type M (multicomponent), Grade NS (nonsag or gunnable) sealant when tested in accordance with Method C 603 meets the requirement for 3 h application life if the time required to empty the cartridge does not exceed 45 s.

7.4 *Hardness*:

7.4.1 Use T (traffic) sealant, specified for use in joints of walkways, plazas, and decks shall have a hardness reading, after being properly cured, of not less than 25 or more than 50 when tested in accordance with Method C 661.

7.4.2 Use NT (nontraffic) sealant, specified for use in nontraffic joints, shall have a hardness reading, after being properly cured, of not less than 15 or more than 50 when tested in accordance with Method C 661.

7.5 *Effects of Heat Aging*—The sealant shall not lose more than 10% of its original weight or show any cracking or chalking when tested in accordance with Method C 792.

7.6 *Tack-Free Time*—There shall be no transfer of the sealant to the plastic film when tested in accordance with Method C 679.

7.7 *Stain and Color Change*—The sealant shall not cause any visible stain on the top surface of a white cement mortar base and shall not itself show a degree of color change that is unacceptable to the purchaser when tested in accordance with Method C 510.

7.8 *Adhesion and Cohesion Under Cyclic Movement*—The total loss in bond and cohesion areas among the three specimens tested for each surface shall be no more than 9 cm² (1½ in.²) when tested in accordance with C 719 with standard mortar, glass, and aluminum or any other specified substrates.

7.9 *Adhesion-in-Peel*—The peel strength for each individual test shall not be less than 2.3 kg (5 lb) when tested in accordance with Method C 794 with standard mortar, glass, and aluminum or any other specified substrate. In addition, the sealant shall show no more than 25% bond loss for each individual test.

7.10 *Adhesion-in-Peel After Ultraviolet Exposure Through Glass*—The peel strength for each individual test shall not be less than 2.3 kg (5 lb) and the compound shall be no more than 25% bond loss for each individual test when tested in accordance with Method C 794.

7.11 *Effects of Accelerated Weathering*—The sealant shall show no cracks greater than those shown in Example #2 of Fig. 1 after the specified ultraviolet exposure and shall show no cracks greater than those shown in Example #2 of Fig. 2 after exposure at cold temperature and the bend test when tested in accordance with Method C 793.

8.   Test Methods

8.1 *Standard Conditions for Laboratory Tests*—All tests described in the following paragraphs shall be performed in a laboratory controlled at 23 ± 2°C (73.4 ± 3.6°F) and 50 ± 5% relative humidity. The sealant sample shall be conditioned at this temperature and relative humidity for at least 24 h before laboratory tests are made.

8.2 *Rheological Properties*—Method C 639.

8.3 *Extrusion Rate*—Method C 603.

8.4 *Application Life*—Method C 603.

8.5 *Hardness*—Method C 661.

8.6 *Effects of Heat Aging*—Method C 792.

8.7 *Tack-Free Time*—Method C 679.

8.8 *Stain and Color Change*—Method C 510.

8.9 *Adhesion and Cohesion After Cyclic Movement*—Method C 719.

8.10 *Adhesion-in-Peel*—Method C 794.

8.11 *Adhesion-in-Peel After Ultraviolet Exposure Through Glass*—Method C 794.

8.12 *Effects of Accelerated Weathering*—Method C 793.

9.   Packaging and Marking

9.1 Packaged materials that are certified by the manufacturer to be in compliance with this specification shall be labeled as to type, class, grade, and use, in accordance with Section 3.

Appendix 2

Recommendations for Upgrading ASTM C-920 Standard

The test methods used in C-920 are satisfactory, but the test exposures and limits have been tightened in some parts of the specification to protect the user. Sealants are available that will meet these more rigid requirements, and specifiers should adopt these recommendations in order to expect improved performance.

Test Method	C-920 Limits	Proposed Limits	ASTM Test Methods
1. Instantaneous Shore A, non-traffic			
(a) after initial cure	25–50	15–40	C-661
(b) after 6 weeks at 158°F	none	15–40	C-661
2. Instantaneous Shore A traffic			
(a) after initial cure	25–60	25–45	C-661
(b) after 6 weeks at 158°F	none	25–45	C-661
3. % weight loss after heat aging			
(a) after 3 weeks at 158°F	10%		C-792
(b) after 6 weeks at 158° F	none	6%	C-792
4. Bond and cohesion area of failure			
(a) after initial cure plus 7 days in water	9cm²		C-719
(b) after initial cure plus 14 days in water	none	9cm²	C-719
(c) after initial cure plus 3 weeks at 158°F plus 2 weeks in water	none	9cm²	C-719
5. Adhesion-in-peel, lbs/inch			
(a) after initial cure plus 7 days in water	5		C-794
(b) after initial cure plus 14 days in water	none	5	C-794

Appendix 3

Recommended Standard for Solvent-Based Acrylic and Chlorosulfonated Polyethylene Sealants (Butyl Excluded)

Test	Requirements	ASTM Test Method
Maximum movement	±12.5%	
Heat-aged hardness maximum (after 3 weeks at 150°F)		
Shore A instantaneous	55	C-661
Adhesion-in-peel, pli	5	C-794
% weight loss, maximum	15	C-792
UV adhesion to glass	5	C-793
Flexibility at −10°F	no failure	C-711
Bubble test	no failure	C-712
Bond cohesion at ±12.5%	no failure	C-910
Recommended practice		C-804

Appendix 4

Recommended Standard for a Butyl Sealant

Test	Requirements	ASTM Test Method
Maximum movement	±7.5%	
Heat-aged hardness after 3 weeks at 158°F, maximum	40	C-661
Adhesion in peel, pli	4 minimum	C-794
UV adhesion, pli	4 minimum	C-794
% weight loss, maximum	25%	C-792
UV/−10°F flex	no failure	C-710
Flexability at −10°F	no failure	C-711
Bubble test	no failure	C-712
Bond cohesion	no failure at ±7.5%	C-910

Appendix 5

Requirements for Latex Sealing Compounds
(ASTM C-834)

Test	Requirements	ASTM Test Method
Extrudability	2 g/s minimum	C-731
Artificial weathering		C-732
Wash-out	none, after aging	
Slump	none, after aging	
Cracking	none, after aging	
Discoloration	none, after aging	
Adhesion loss	25% maximum	
Volume shrinkage	30% maximum	C-733
Low-temp. flexibility		
at 0°F	no adhesion loss	C-734
Recovery	75% minimum	C-736
Adhesion loss	25% maximum	C-735
Slump	4mm maximum	D-2202
Stain index	3 maximum	D-2203
Tack-free time	no material stick after 72 hours	D-2377

Appendix 6

Requirements for Oil-and-Resin-Based Caulking Compound for Building Construction (ASTM C-570)

Property	Requirements	ASTM Method
Shrinkage, maximum	20%	D-2453
Tenacity, minimum folds	6	D-2453
Bond, maximum % loss	10	D-2450
Slump, maximum	0.15 inch	D-2202
Stain, maximum index	6.0	D-2202
Tack-free time	5 hours maximum	D-2377
Extrudability	9 s/milli-liter maximum	D-2452
Lead content	1% maximum on total solids	D-2088

Appendix 7

Requirements for Glazing Compounds for Back Bedding and Face Glazing of Metal Sash (ASTM C-669)

Property	Requirements	Test Method
Degree of set on cured compound	2 to 12 mm penetrometer	ASTM D-2451
Accelerated weathering tests		
Surface cracking and peeling	no. 5 max. of plate no. 1	Fed. Spec. TT-G-410E
Deep bead cracking	no. 5 max. of plate no. 2	Fed. Spec. TT-G-410E
Adhesion loss	no. 5 max. of plate no. 3	Fed. Spec. TT-G-410E
Wrinkling	no. 5 max. of plate no. 4	Fed. Spec. TT-G-410E
Oil exudation	no. 5 max. of plate no. 5	Fed. Spec. TT-G-410E
Slump	none	ASTM C-2376

Appendix 8

Recommended Specification Limits for Glazing Tapes

Test	Texture of Tapes		ASTM Test Method
	Firm	Soft	
Weight loss on aging, % maximum	3	3	C-771
Low-temp. flexibility at 158°F/−10°F	no failure	no failure	C-765
Adhesion loss	50% maximum	50% maximum	C-766
Oil migration[a]	none	none	C-772
Softness, minimum	5.0–6.0	9.3–11.5	C-782
Tensile adhesion	16–18 psi	15–17 psi	C-907
Yield strength, psi	14–16 psi	11–13 psi	C-908
elongation %	700–900	1300–1500	C-908
Release paper test	no adhesion	no adhesion	C-879

[a]Material for use with cap bead should show no migration. Some migration may be desired for wood sash and masonry surfaces for better wetting.

329

Bibliography

1. J. R. Panek, "Building Seals and Sealants," ASTM Special Technical Publication-606, 1976.

2. Per Gunnar Burstrom, "Ageing and Deformation Properties of Building Materials," Report TVBM-1002, Division of Building Materials, University of Lund, Sweden, 1979.

3. W. A. Oberdick, "Computerized Joint Movement Criteria," Proceedings, Midyear Technical Seminar, ASTM Committee C-24, 1968.

4. N. V. Morosov, "Joints of Large Panel Building and their Characteristics," NBRI Report 51C, Norwegian Building Research Institute, Oslo, 1968.

5. M. Dalaker, "Gaskets in Window Joints," NBRI Report 51C, Norwegian Building Research Institute, Oslo, 1968.

6. J. P. Cook, "A Study of Polysulfide Sealants for Joints in Bridges," Highway Research Board Bulletin 299, 1965.

7. H. R. Brown, "Polychloroprene Gaskets," in A. Damusis (Ed.), Sealants, New York: Reinhold, 1967.

8. W. Koppes, "Functional Requirements, Standards and Tests for Joint Seals," NBRI Report 51C, Norwegian Building Research Institute, Oslo, 1968.

9. K. K. Karpati, "Development of Test Procedure for Predicting Performance of Sealants," American Chemical Society Symposium Series No. 113, "Plastic Mortars, Sealants, and Caulking Compounds," 1979.

10. K. K. Karpati, "Weathering of Silicone Sealant on Strain-Cycling Exposure Rack," Adhesives Age, 23, 11 (November 1980).

11. G. K. Garden, "Sensible Use of Sealants," NBRI Report 51C, Norwegian Building Research Institute, Oslo, 1968.

12. J. R. Panek, "Applications for Polysulfide Polymers," in Polyethers Part III, New York: Interscience, 1962.

13. R. J. Boot, "Silicone Sealants," in A. Damusis (Ed.), Sealants, New York: Reinhold, 1967.

14. K. K. Karpati, K. R. Solvason, and P. J. Sereda, "Weathering Rack for Sealants," Journal of Coatings Technology, 49, 626 (March 1977).

15. A. Damusis, "Urethane Sealants," in A. Damusis (Ed.), Sealants, New York: Reinhold, 1967.

16. R. M. Evans, and R. B. Greene, "Urethane Sealants," in J. Panek (Ed.), Building Seals and Sealants, ASTM Special Technical Publication-606, 1976.

17. J. J. Higgins and N. E. Stucker, "Butyl Rubber and Polyisobutylene," in I. Skeist (Ed.), *Handbook of Adhesives*, 2nd Ed., New York: Van Nostrand-Reinhold, 1977.

18. R. S. Yanasaki, "Coatings to Protect Concrete Against Damage by De-Icer Chemicals, a Literature Review," *Journal of Paint Technology*, **9**, 509 (June 1967).

19. J. A. Dalton and P. V. Paulus, "Lock-Strip Glazing Gaskets," in J. Panek (Ed.), *Building Seals and Sealants*, ASTM—STP-606, 1976.

20. A. H. Smith Jr., "Cast Iron Soil Pipe Gaskets," in J. Panek (Ed.), *Building Seals and Sealants*, STP-606, 1976.

21. J. P. Cook and R. M. Lewis, "Evaluation of Pavement Joint and Crack Sealing Materials and Practices," NCHRP Report No. 38, Highway Research Board, 1967.

22. K. A. Stead, "Control of Pavement Movement Adjacent to Structures," Report No. 1, Engineering Experimental Station, University of Mississippi, 1965.

23. I. Minkarah, J. P. Cook, J. F. McDonough, and S. Jaghoory, "Effect of Different Variables in Horizontal Movement of Concrete Pavement," Publication SP-70, American Concrete Institute, 1980.

24. G. A. Busch, "Experience with Elastomeric Concrete Expansion Joint Transition Dam in Bridges," Publication SP-70, American Concrete Institute, 1980.

25. S. Spells and J. M. Klosowski, "Silicone Sealants for Use in Concrete Construction," Publication SP-70, American Concrete Institute, 1980.

26. F. D. Gaus, "Essential Elements for High Performance Joint Sealing and Resealing Portland Cement Concrete Pavements," Publication SP-70, American Concrete Institute, 1980.

27. C. Seibel, Jr., "Development and Field Performance of New ASTM Standards in Hot-Applied, Formed-in-Place, Joint Sealants for Highway and Airfield Pavements," Publication SP-70, American Concrete Institute, 1980.

28. M. Emerson, "Thermal Movements of Concrete Bridges, Field Measurements and Methods of Prediction," Publication SP-70, American Concrete Institute, 1980.

29. G. S. Puccio, "Extruded Seals for Bridges and Structures," Publication SP-70, American Concrete Institute, 1980.

30. S. C. Watson, "Sealing of Airfield and Highway Pavements Using Preformed Extruded Compression Joint Seals," Publication SP-70, American Concrete Institute, 1980.

31. F. W. Reinhart and I. G. Calloman, "Survey of Adhesion and Adhesives," WADC Technical Report 58-450, 1959.

32. I. Skeist, "Handbook of Adhesives," 2nd Ed., New York: Van Nostrand-Reinhold, 1976.

33. R. B. Rohrer, "Adhesives for Floor Surfacing Materials," in *Adhesives and Sealants in Buildings*, No. 577, Washington, D.C.: Building Research Institute, 1959.

34. L. I. Knab, "Polyester Concrete, Load Rate Variance," Highway Research Record No. 287, Highway Research Board, 1969.

35. R. T. Howe, "Polyvinyl Acetate and Portland Cement Mortars," *Journal of the Construction Division*, Proceedings of the American Society of Civil Engineers, Paper no. 2358, 1960.

36. D. J. McKee and J. P. Cook, "A Study of Adhesive Bonded Concrete — Metal Deck Slabs," Transportation Research Record No. 762, Transportation Research Board, 1981.

37. M. R. Toensmeyer and J. P. Cook, "Repair of Torsionally Inadequate Concrete Beams Using Bonded Steel Plates," Transportation Research Record No. 762, Transportation Research Board, 1981.

38. C. A. K. Irwin, "Strengthening of Concrete Beams by Bonded Steel Plates," TRRL Supplementary Report 160VC, Department of the Environment, Transportation Road Research Laboratory, United Kingdom, 1975.

39. J. P. Cook, *Composite Construction Methods*, New York: Wiley, 1977.

*Glossary**

ABRASION RESISTANCE.   Resistance to wear resulting from mechanical action of a surface.

ACCELERATED AGING.   A set of laboratory conditions designed to produce in a short time the results of normal aging. Usual factors included are temperature, light, oxygen, and water.

ACCELERATOR.   An ingredient used in small amounts to speed up the action of a curing agent. Sometimes used as a synonym for curing agent.

ACETONE.   Dimethyl ketone. A very volatile solvent. Particularly useful for cleaning metal substrates.

ACRYLIC.   A group of thermoplastic resins or polymers formed from the esters of acrylic acid.

ACTIVATOR.   A material that, when added to a compound or curing agent, will speed up the curing mechanism.

ADHEREND.   A body that is held to another body by an adhesive.

ADHESION.   The clinging or sticking together of two surfaces. The state in which two surfaces are held together by forces at the interface.

ADHESION, MECHANICAL.   Adhesion due to the physical interlocking of the adhesive with the surface irregularities of the substrate.

*Sources used in preparation of this glossary:

American Concrete Institute, Committee Report 504.
C. M. Gay, and H. Parker, "Materials and Methods of Architectural Construction," Wiley, 1943.
Highway Research Board, Committee MC-D3, "Glossary of Sealant Terminology."
W. C. Huntington, "Building Construction," Wiley, 1964.
Flat Glass Marketing Association, "Glazing Manual," 1974.
Flat Glass Marketing Association, "Glazing Sealing Systems Manual," 1970.

ADHESION, SPECIFIC. Adhesion due to molecular forces at the surface.

ADHESIVE. A substance capable of holding materials together by surface attachment.

ADHESIVE FAILURE. Type of failure characterized by pulling the adhesive or sealant loose from the adherend.

ADSORPTION. The action of a body in condensing and holding gases and other materials at its surface.

AGING. The progressive change in the chemical and physical properties of a sealant or adhesive.

ALLIGATORING. Cracking of a surface into segments so that it resembles the hide of an alligator.

AMBIENT TEMPERATURE. Temperature of the air surrounding the object under construction.

AROCLOR. A chlorinated diphenyl that has been used as a plasticizer in some sealants. It has been banned by OSHA.

ASBESTOS. A mineral with a structure of long, fine fibers.

ASPHALT. Naturally occurring mineral pitch or bitumen.

BACK-UP. A compressible material used at the base of a joint opening to provide the proper shape factor in a sealant. This material can also act as a bond-breaker.

BEAD. A sealant or compound after application in a joint irrespective of the method of application, such as caulking bead, glazing bead, and so on.

BEDDING COMPOUNDS. Any material into which another material such as a plate of glass or a panel, may be embedded for close fit.

BIREFRINGENCE. The refraction of light in two slightly different directions to form two rays. The phenomenon can be used to locate stresses in a transparent material.

BITE. Amount of overlap between the stop and the panel or light.

BLOWN OILS. Oils that have had air blown through them to increase viscosity or alter other properties.

BOND (NOUN). The attachment of an interface between substrate and adhesive, or sealant.

BOND (VERB). To join materials together using an adhesive.

BOND-BREAKER. Thin layer of material such as tape used to prevent the sealant from bonding to the bottom of the joint.

BOND DURABILITY. A test cycle in ASTM C-920 for measuring the bond strength after repeated weather and extension cycling.

BOND FACE. The part or surface of a building component that serves as a substrate for an adhesive or sealant.

BOND STRENGTH. The force per unit area necessary to rupture a bond.

BULK COMPOUNDS. Any sealant or caulk that has no definite shape and is stored in a container or cartridge.

BUTT JOINT. A joint in which the structural units are joined to place the adhesive or sealant into tension or compression.

BUTYL RUBBER. A copolymer of essentially isobutene with small amounts of isoprene. As a sealant it has low recovery and slow cure.

CARBON BLACK. Finely divided carbon formed by the incomplete combustion of natural gas. May be used as a reinforcing filler in sealants.

CATALYST. Substance added in small quantities to promote a reaction, while remaining unchanged itself. Sometimes referred to as the curing agent for sealants.

CAULK (NOUN). A material with a relatively low movement capability, usually less than ±10%. Generally refers to oil-based caulks, and sometimes to butyl and acrylic latex caulks. ASTM C-24 proposes that all materials be termed sealants.

CAULK (VERB). To fill the joints in a building with a material.

CELLULAR MATERIAL. A foamy material containing many small cells dispersed throughout the material. The cells may be either open or closed. Density is usually described in terms of pounds per cubic foot.

CHAIN STOPPER. A material added during the polymerization process to terminate or control the degree of reaction. This could result in soft sealants, or higher elongation.

CHALKING. Formation of a powdery surface due to weathering.

CHECKING. The formation of slight breaks or cracks in the surface of a sealant.

CHEMICAL CURE. Curing by chemical reaction. Usually involves the cross-linking of a polymer.

CLOSED-CELL FOAM. A foam that will not absorb water because all the cells have complete walls.

COEFFICIENT OF EXPANSION. The coefficient of linear expansion is the ratio of the change in length per degree to the length at 0°C.

COHESION. The molecular attraction that holds the body of a sealant or adhesive together. The internal strength of an adhesive or sealant.

COHESIVE FAILURE. The failure characterized by pulling the body of a sealant or adhesive apart.

COMPATIBLE. Two or more substances that can be mixed or blended together without separating, reacting, or affecting the material adversely. However, two materials such as a sealant and a tape gasket are compatible if there is no interaction between them and materials from one do not migrate into the other.

COMPRESSION SEAL.   A preformed seal that is installed by being compressed and inserted into the joint.

COMPRESSION SET.   The amount of permanent set that remains in a specimen after removal of a compression load.

CONE PENETROMETER.   An instrument for measuring the relative hardness of soft deformable materials.

CRAZING.   A series of fine cracks that may extend through the body of a layer of sealant or adhesive.

CREEP.   The deformation of a body with time under constant load.

CROSS-LINKED.   Molecules that are joined side-by-side as well as end-to-end.

CURE.   To set up or harden by means of a chemical reaction.

CURE TIME.   Time required to effect a cure at a given temperature.

CURING AGENT.   A chemical that is added to effect a cure in a polymer or sealant.

DEPOLYMERIZATION.   Separation of a complex molecule into simpler molecules. Also softening of a sealant by the same action.

DOWEL.   A straight steel bar used to transfer a load between sections of a concrete pavement slab.

DUROMETER.   An instrument used to measure hardness or Shore hardness. Also may refer to the hardness rather than the instrument.

ELASTICITY.   The ability of a material to return to its original shape after removal of a load.

ELASTOMER.   A rubbery material that returns to approximately its original dimensions in a short time after a relatively large amount of deformation.

EMULSION.   A suspension of microscopic particles in water.

EPOXY.   A resin formed by combining epichlorohydrin and bisphenols. Requires a curing agent for conversion to a plastic-like solid. Has outstanding adhesion and excellent chemical resistance.

EXOTHERMIC.   A chemical reaction that gives off heat.

EXTENDER.   An organic material used as a substitute for part of the polymer to lower the cost of a sealant or adhesive.

EXTENSIBILITY.   The ability of a sealant to stretch under tensile load.

EXTRUSION FAILURE.   Failure that occurs when a sealant is forced too far out of the joint. The sealant may be abraded by dirt or folded over by traffic.

FATIGUE FAILURE.   Failure of a material due to rapid cyclic deformation.

FIELD-MOLDED SEALANT.   A bulk compound that takes its shape by being placed into a joint.

FILLER.   Finely ground material added to a sealant or adhesive to change or improve certain properties. If used to excess it cheapens the compound.

FLASHING.   Strips, usually of sheet metal or rubber, used to waterproof the junctions of building surfaces, such as roof peaks and valleys, and the junction of a roof and chimney.

GASKET.   A cured elastic but deformable material placed between two surfaces to seal the union between the surfaces.

GUNABILITY.   The ability of a sealant to extrude out of a cartridge in a caulking gun.

GYPSUM WALLBOARD.   A sandwich-type material. Gypsum plaster with a heavy paper coating on both sides. When fastened directly to studs, it forms a wall surface.

HARDBOARD.   Fine pieces of wood bound together with an adhesive and pressed into 4 × 8-foot sheets. Thicknesses are approximately ⅛, ³⁄₁₆, and ¼ inches. Thermosetting resins are usually used as the adhesive binder.

HARDENER.   A substance added to control the reaction of a curing agent in a sealant or adhesive. Sometimes used as a synonym for curing agent or catalyst.

HARDNESS.   The resistance of a material to indentation. On the Shore A Durometer Scale, hardness is measured in relative numbers from 0 to 100 for rubber-like materials.

HEAD.   The top member of a window or door frame.

HEEL BEAD.   Sealant applied at the base of channel. After setting light or panel and before the removable stop is installed, one of its purposes is to prevent leakage past the stop. The sealant must bridge the gap between the glass and frame.

HOCHMAN TEST CYCLE.   The bond durability test cycle used in ASTM C-290.

HYPALON.   A chlorosulfonated polyethylene synthetic that has been used as a base for making solvent-based sealants.

INTERFACE.   The common boundary surface between two substances.

JAMB.   The side of a window, door opening, or frame.

JOINT (ADHESIVE USE).   The point at which two substrates are joined by an adhesive.

JOINT (SEALANT USE).   The opening between component parts of a structure.

LAITANCE.   A thin, weak coating that sometimes forms on the surface of concrete, and is caused by water migration to the surface.

LAP JOINT.   A joint in which the component parts overlap so that the sealant or adhesive is placed into shear action.

LATEX.   A colloidal dispersion of a rubber resin (synthetic or natural) in water, which coagulates on exposure to air.

LATEX CAULKS.   A caulking material made using latex as the raw material. The most common latex caulks are polyvinyl acetate or vinyl acrylic.

LOAD-TRANSFER DEVICE.   Any device embedded in the concrete on both sides of a pavement joint to prevent relative vertical movement of slab edges.

MASTIC.   A thick, pasty coating.

MERCAPTAN.   An organic compound containing –SH groups.

MIL.   One-thousandth of an inch.

MODULUS.   The ratio of stress to strain. Also the tensile strength at a given elongation.

MONOMER.   A material composed of single types of molecules. A building block in the manufacture of polymers.

MULLION.   External structural member in a curtain-wall building. Usually vertical. May be placed between two opaque panels, between two window frames, or between a panel and a window frame.

MVT.   A measurement of the moisture vapor transmission through a film, usually expressed in terms of grams of water per square meter per 24 hours.

NECK DOWN.   The change in the cross-sectional area of a sealant as it is extended.

NEEDLE GLAZING.   The application of a small bead of sealant using a nozzle not exceeding ¼ inch in diameter.

OIL (DRYING).   Oils that dry to a hard varnish-like film. Linseed oil is a common example.

OPEN-CELL FOAM.   A foam that will absorb water and air because the walls are not complete and run together.

OPEN TIME.   Time interval from when an adhesive is applied to when it becomes unworkable.

OXIDATION.   Formation of an oxide. Also the deterioration of rubbery materials due to the action of oxygen or ozone.

OZONE.   A reactive form of oxygen. A powerful oxidizing agent, it occurs in the atmosphere.

PARTICLE BOARD.   Same as hardboard except that larger wood chips are used as the filler.

PAVEMENT GROWTH.   An increase in the length of a pavement caused by incompressibles working into the joints.

PEEL TEST.   A test of an adhesive or sealant using one rigid and one flexible substrate. The flexible material is folded back (usually 180°) and the substrates are peeled apart. Strength is measured in pounds per inch of width.

PERMANENT SET.   The amount of deformation that remains in a sealant or adhesive after removal of a load.

PHENOLIC RESIN.   A thermosetting resin. Usually formed by the reaction of a phenol with formaldehyde.

PIGMENT.   A coloring agent added to a sealant.

PITCH.   The residue that remains after the distillation of oil and other substances from raw petroleum.

PLASTICIZER.   A material that softens a sealant or adhesive by solvent action, but is relatively permanent.

PLASTISOL.   A physical mixture of resin (usually vinyl) compatible plasticizers, stabilizers, and pigments. Mixture requires fusion at elevated temperatures in order to convert the plastisol to a homogenous plastic material.

POISE.   The cgs unit of viscosity. Example: A polysulfide highway joint sealant might have a viscosity of 500 poises, at 77°F. Higher numbers indicate a more viscous material.

POLYBUTENE BASE.   Compound made from polybutene polymers.

POLYESTER.   Resins manufactured by reacting a dicarboxylic acid and a dihydroxy alcohol. Polyesters are used in one-part and two-part systems for coatings, molding compounds, and the manufacture of dacron, which is a polyester fiber.

POLYETHYLENE.   A straight-chain plastic polymer of ethylene used for containers, packaging, and the like.

POLYMER.   A compound consisting of long chain-like molecules. The building units in the chain of polymers.

POLYSULFIDE RUBBER.   Synthetic polymer usually made using sodium polysulfide. Also polymers that contain mercaptan terminals that when cured form disulfide linkages.

POT LIFE.   See Working life.

PREFORMED SEALANT.   A sealant that is preshaped by the manufacturer before being shipped to the job site.

PRESSURE-SENSITIVE ADHESIVE.   Adhesive that retains tack after release of the solvent so that it can be bonded by simple hand pressure.

PRIMER.   In building construction, a compatible coating designed to enhance adhesion.

REFLECTION CRACKS.   A crack through a bituminous overlay on portland cement concrete pavement. The crack occurs above any working joint in the base pavement.

REINFORCEMENT.   In rubbers this is the increase in modulus, toughness, tensile strength, and so forth, by the addition of selected fillers.

RESILIENCE.   A measure of energy stored and recovered during a loading cycle. It is expressed in percent.

RESINS.   Solid or liquid organic materials, generally not soluble in water, that have little or no tendency to crystallize. Example: epoxy and polyester resins.

RETARDER.   A substance added to slow down the cure rate of a sealant or adhesive. Mild acids are retarders for polysulfide sealants.

ROUTING.   Removing old sealant from a joint by means of a rotating bit.

RUBBER LATEX.   Water emulsion of an elastomer.

SEAL (NOUN).   ASTM definition is "A material applied in a joint or on a surface to prevent the passage of liquids, solids, or gases."

SEALANT.   ASTM definition is "In building construction, a material which has the adhesive and cohesive properties to form a seal." Sometimes defined as an elastomeric material with a movement capability greater than $\pm 10\%$.

SEALANT BACKING.   In building construction, a compressible material placed in a joint before applying a sealant.

SEALER.   A surface coating generally applied to fill cracks, pores, or voids in the surface.

SEALING TAPE.   ASTM definition is "A preformed, uncured or partially cured material which when placed in a joint, has the necessary adhesive and cohesive properties to form a seal."

SELF-LEVELING SEALANT.   A sealant that is fluid enough to be poured into horizontal joints. It forms a smooth, level surface without tooling.

SHAPE FACTOR.   The width-to-depth proportions of a field-molded sealant.

SHEAR TEST.   A method of deforming a sealed or bonded joint by forcing the substrates to slide over each other. Shear strength is reported in units of force per unit area (psi).

SHELF LIFE.   The length of time a sealant or adhesive can be stored under specific conditions and still maintain its properties.

SHORE "A" HARDNESS.   The measure of firmness of a rubbery compound or sealant by means of a Durometer Hardness Gauge. An art gum eraser has an approximate Shore A hardness of 20 to 25. A rubber heel might be in the 80 to 90 range.

SHRINKAGE.   Percentage weight loss of volume loss under specified accelerated conditions.

SILICONE RUBBER.   A synthetic rubber based on silicon, carbon, oxygen, and hydrogen. Silicone rubbers are widely used as sealants and coatings.

SKEWED JOINTS.   Transverse joints in a pavement slab, which are placed at an angle and not perpendicular to the direction of traffic.

SOLVENT.   Liquid in which another substance can be dissolved.

SPALLING.   A surface failure of concrete, usually occurring at the joint. It may be caused by incompressibles in the joint, by overworking the concrete, or by sawing joints too soon.

SPONGE.   Cellular material, usually of open-cell construction.

STOPLESS GLAZING.   The use of a sealant as a glass adhesive to keep glass in permanent position without the use of exterior stops.

STRAIN.   Deformation per unit length. Example: Change in length divided by the original length of a test specimen. Strain is a dimensionless number. Small strains are expressed in units of inches per inch. Strains in rubbery materials may be expressed in percent.

STRESS.   Force per unit area, usually expressed in pounds per square inch (psi).

STRESS RELAXATION.   Reduction in stress in a material that is held at a constant deformation for an extended time.

STRUCTURAL GLAZING GASKETS.   A synthetic rubber section designed to engage the edge of glass or other sheet material in a surrounding frame by forcing an interlocking filler strip into a grooved recess in the face of the gasket.

STRUCTURAL SEALANT.   A sealant used as an adhesive to bond two materials together, as in stopless glazing.

SUBGRADE.   The earth or granular fill below a pavement slab.

SUBSTRATE.   An adherend. The surface to which a sealant or adhesive is bonded.

TACKINESS.   The stickiness of the surface of a sealant or adhesive.

TAPE.   See Sealing Tape.

TEAR STRENGTH.   The load required to tear apart a sealant specimen. ASTM test method D-624 expresses tear strength in pounds per inch of width.

TENSILE STRENGTH.   Resistance of a material to a tensile force (a stretch). The cohesive strength of a material expressed in psi.

THERMOPLASTIC.   A material that can be repeatedly softened by heating. Thermoplastics generally have little or no chemical cross-linking.

THERMOSETTING.   A material that hardens by chemical reaction. Nor remeltable. The reaction usually gives off heat which is noticeable if the reaction is rapid.

THIXOTROPIC. Nonsagging. A material that maintains shape unless agitated. A thixotropic sealant can be placed in a joint in a vertical wall and will maintain its shape or position without sagging during the curing process.

TOOLING. The act of compacting and shaping a material in a joint.

TOXIC. Poisonous or dangerous to humans by swallowing, inhalation, or contact resulting in eye or skin irritation.

TRANSVERSE JOINT. A joint perpendicular to the direction of traffic in a highway pavement.

ULTIMATE ELONGATION. Elongation at failure.

ULTRAVIOLET LIGHT (UV). Part of the light spectrum. Ultraviolet rays can cause chemical changes in rubbery materials.

UNITED INCHES. The total of one width and one height of a light of glass in inches.

URETHANE. A family of polymers ranging from rubbery to brittle. Usually formed by the reaction of a diisocyanate with a hydroxl compound with or without amines or amides. This is a major class of sealants.

VEHICLE. The liquid component of a material. Example: oil paint is composed of a vehicle (linseed oil) and pigment.

VISCOSITY. A measure of the flow properties of a liquid or paste. Example: honey is more viscous than water. Water (the standard of comparison) has a viscosity of $1/100$ of a poise or 1 centipoise at a specified temperature.

VULCANIZATION. Improving the elastic properties of a rubber by chemical change, usually heat.

WEATHEROMETER. An environmental chamber in which specimens are subjected to water spray and ultraviolet light.

WORKING LIFE. Period of time after mixing, during which a sealant or adhesive can be used.

Index